Biography of AN ISLAND

BOOKS BY PERRY D. WESTBROOK

ACRES OF FLINT: WRITERS OF RURAL NEW ENGLAND, 1870–1900

BIOGRAPHY OF AN ISLAND

Biography of AN ISLAND

by

Perry D. Westbrook

South Brunswick
New York • Thomas Yoseloff • London

Library of Congress Catalogue Card Number: 57-12360

Thomas Yoseloff, Publisher
Cranbury, New Jersey 08512

Thomas Yoseloff Ltd
108 New Bond Street
London W1Y OQX, England

SBN: 498 07005 0
Printed in the United States of America

To

CLYDE TORREY

Preface

IN THE HUGE ACCUMULATION OF WRITING ABOUT THE MAINE coast I have been vain enough to think there is room for one more book which avoids the excesses of both the sentimentalists and the novelists. I say "one more" because a select few —like Celia Thaxter's *Among the Isles of Shoals* and Mary Ellen Chase's *A Goodly Heritage*—are marked by the restraint and objectivity of effective regional writing. Ever since first visiting Swans Island in 1940 I have felt that there was material for a book in this community which through its isolation has retained so much of an earlier New England and America and which in its encystment in the sea exemplifies so well the adaptability of man to his natural environment.

To present the continuing life of this township I have had to cut across many of the formal disciplines. I have had to play the antiquarian, the ecologist, the sociologist, even the literary critic. Insofar as a community has organic existence, I have tried to be the biographer of Swans Island, and as a biographer I have had to assume many roles. But always in presenting to the reader a picture of an interesting and vanishing way of life, I have made accuracy my chief goal.

Whatever accuracy I have been able to achieve is owing to the helpfulness of many people: the islanders themselves, especially the "town fathers," without whose friendliness and good humor this work would not have been possible; and the librarians of the Maine and the Massachusetts Historical Associations, the Maine and New York State Libraries, and the Bowdoin College and the University of Maine Libraries, all of whom aided in digging out the meager but indispensable manuscript and printed material bearing on my subject.

To the John Simon Guggenheim Memorial Foundation I owe the year of freedom that enabled me to experience winter life on the island and to do the actual writing of the book.

Contents

Illustrations

(*The illustrations appear as a group following page 128*)

Biography of
AN ISLAND

1

The Place

SWANS ISLAND LIES FOUR MILES SOUTHWEST OF MT. DESERT, Maine, at the mouth of Bluehill Bay, which bites deeply into the mainland to the north. In the island's narrowest spot, which is a former Indian portage called the Carrying Place, one can throw a stone from water to water at high tide. At its broadest place one would have to thrash through four miles of spruce jungle to cross it. To walk the shoreline would require at least three days of scramble over rocks, along cobble beaches, and up and down ledges and escarpments. The north and south shores form something resembling a segment of a double-edged saw with huge jagged teeth. Narrow, steep-sided promontories jut far out into the sea on the south or into the bay on the north, and between them are the long coves. Like the many ridges and valleys throughout New England, inlets and points run in a north-south direction—the direction in which the glaciers moved.

This tattered shoreline has given the island many anchorages. Burnt Coat Harbor on the south side is the best of them, some say the best small harbor in Eastern Maine. It is over a mile long and a half mile broad with enough water to float all but the largest vessels. An island protects its mouth and projects beyond on either side, leaving two narrow channels at right angles to the harbor. The deeper and wider "thoroughfare" comes from the west. Above it on Hocomock Head, the highest headland of the island, perches a lighthouse. In the stormiest weather this harbor is never rougher than a lake of the same size. The first settlements were on its shore, and today there is a village on either side of it. Other coves on both the north and south shores have their clusters of houses, wharves, and moorings.

To the south, the ocean, sprinkled and festooned with islets, reaches to the horizon. Outer Long Island squats stolidly on the sea to the southeast. To the southwest chains of treeless isles and ledges arch out toward Isle au Haut, whose long black ridge stands like a continent among all these splatterings and driblets of land and rock. The northern promontories, especially Roderick's Head, command a view of the Mt. Desert range across Bluehill Bay. Over the inner islands up the bay is Blue Hill itself. On days of sparkling clearness the rounded summits of the Camden Hills, thirty miles distant on the western shore of Penobscot Bay, bulge above the intervening islands.

The east and west shores, where there are no anchorages, are deserted. Here the spruce woods crowd down to high-water mark. In places the shore is perpendicular rock carved and gouged by the waves into thundering gorges and resonating caves out of which the spray plumes upward to catch the sunlight. One of these, called the Spouting Horn or Benny's Den, spews and roars as lustily as the famous Thunder Hole near Bar Harbor.

The walker has to keep far back in the woods to avoid the treacherous crevices which creep in under the mold. In places the ledges flatten out. Glacial scars seam them from north to south and one wanders as on a pavement among the great boulders left there by waves and prehistoric ice. In the depressions at low tide are pools astir with life, the water warm to feet burning from constant pounding on granite. Here the walker takes his rest, watching the breakers tumble up the sloping rocks.

Such stretches of smooth walking are few. Usually, where it is not precipitous, the shore is a jumble of boulders. If the ledges have broken up comparatively recently, the rocks are rough and angular. Sometimes, where the frost has split them off in layers along the flaws in the granite, they are like quarried blocks. Walking is hazardous. Some of the stones, especially the flatter ones, are so nicely balanced that they tip up under the slightest weight, threatening to "hoist" the unwary, as the islanders say, against sharp edges. But where the waves have had sufficient centuries to work, the raw chunks have been milled into glossy spheres and ovoids of every size. They roll and slide underfoot, growling admonitions of sprained ankles and blistered heels.

Where the shoreline dips, these cobble beaches build up into "sea walls." If the shore had been a few feet lower there would have been a cove. But instead the sea has erected a barrier across the indentation and behind the barrier is either a marsh or a brackish pond that changes its level with the ebb and flow of the tide seeping through the stones. The top of a sea wall is flat, about the width of a city sidewalk, and as high as the highest storm tide. The slope is gentle on the sea side, but toward the land it is very steep, resembling the wall of a fortress. The ocean has deposited its storm-gathered debris along the crest of the wall: heaps of sea-weed, stumps of trees, lengths of cordwood, sections of weirs, shattered lobster traps entangled in their rotting warps, hatches, and timbers from wrecked vessels and barges. Here and there a beach pea has taken root and nods its purple flowers at the walker, or a patch of sea lungwort lays down a carpet of mist-gray leaves. On the land side wild irises and roses bedizen the great timbers tossed over the wall by some mountainous tide of years ago.

Occasionally a ferocious winter storm chews a hole through the wall which the sea has so carefully constructed. But in the ensuing calmer weather the partition is rebuilt. The pond once more has its separate existence, yet is always in phase with the surge and cadence of the power that created it.

A child, a beached fisherman, a summer visitor will sometimes loiter along this ridge of cobble between the pond and the ocean during the day, stooping to pick up and handle the more perfectly shaped stones, worn to glassy smoothness by their interminable rotation up and down the beach, of every hue—emerald, turquoise, ruby, jet, cobalt. The breeze, always fresh here in summer, will blow in his face as he watches sails tilted against the horizon and gulls riding the air currents over the bay. The glare, the color, will pierce his eyes finally with pain, and he will seek the cool darkness of the spruce woods nearby.

Perhaps at the end of the wall a fishing shack teeters just under an embankment down which a footpath slithers from a blueberry pasture. The windows of the shack have long been glassless, and the doors have disappeared in ancient hurricanes. A few planks hang slantingly from half-sprung nails on weathered studdings. The roof, supported by an intricate understructure of spruce poles, lets in the weather through its sagging boards, its only function being to provide shade for the bait stored inside. From the distance the shack looks like a heap of well-bleached driftwood tossed there by some freakish combination of tide and wind. So harmoniously does it blend with the beach that it might first be detected by its smell—the stench of fish scales accumulated an inch deep on the floors, of bait barrels empty of everything except their asphyxiating breath, of sun-baked tar, and of herring rotting in forgotten trawl tubs. Half-way across the sea wall the fish house aura reaches out to the straggler and lures him on; for only at close quarters, at the very mouth of the bait-barrel or in the farthest corner of a well-built, air-tight bait house—and such exist—is the odor unpleasant. From a distance, diluted by the salt breeze and the odor of spruces, it is the essence of the excitement of the sea.

2

The island tides are the violent ones of the whole eastern coast of Maine. The mean interval between high and low is nearly eleven feet. But the spring tides, those that come with the full and new moons, add three or four more feet. During their ebb and flow, swirling currents sweep the shores of the island. To the east, the water of half the bay races by, dragging the trap buoys beneath the surface. Anchoring is impossible, and the handliner, even with a three-pound sinker, can't get his hook to the bottom. The water laps against the tops of the wharves and creeps to within inches of the roads. Creeks and marshes far back from the shore gurgle into life. Logs of driftwood stranded two weeks before at the edge of the woods or meadows float again and, if the wind is offshore, vanish in search of some new resting place. The waves lap among the grass and beach goldenrod where the water has no right to be. These spring tides accomplish a bi-weekly cleansing of the shore. Decaying seaweed, old paper, and bottles are swept away. If a storm coincides with the high tides, the roads and wharves are awash. Structures collapse and equipment disappears. Embankments are undermined and soil is lost to the sea.

The low tides which occur at these times are even more amazing than the high. The islanders call them "low drain tides," and they do threaten to drain the ocean to the bottom. The shores rise higher, more forbiddingly from the water. The ledges and headland cliffs don low-fitting skirts of rockweed. The sound of the waves recedes till it is no more than a reminiscence. Rocks, remnants of old wharves, sunken timbers that appear at no other time are revealed by the shrinking ocean.

The tide must be consulted in every activity of island life. To the fishermen—particularly the weir operators, the seiners, the inshore trawlers—the tides are despots. But their sway reaches out into the commonplaces of life as well. Is a boat to be painted, a fisherman's meal to be gotten, a swim to be taken? One must consult the tide before acting. People leaving or arriving try to time their passage with high tide, especially if they have much luggage, because then they can step directly off a wharf into a boat, or the boat can nose right up into the sand at the head of a cove with the assurance that in ten minutes it will float free on the rising water. Embarkation at low tide means dizzying climbs down slippery ladders on the side of wharves and the use of hoists to load and unload baggage; or where no wharf is available, complicated maneuvers with punts. If a fisherman wishes to make repairs to the bottom of his workboat, he can run it as far up the beach as he can at high water. An hour later it will be "grounded out" and he will have ten hours of working time before his boat floats again. The landsman's almost sacred regularity of meals and bed and rising time mean nothing to many of the fishermen, particularly the older ones. Mainlanders of fixed habit are shocked when some old fisherman, beached for many years, still has his supper in the middle of the morning, arises one day at two A.M. and the next at two P.M., or gets up in the middle of the night for four hours of work and then goes back to bed again.

So tyrannical is the tide's control of the islander's life that it is little wonder he imagines it extends into other areas, notably that of the weather. Long and short range forecasting, generally inaccurate, is based on the movements of the tide. Is there a heavy fog? It will "scale off" with the turn of the tide. Have we been enjoying a week of exceptionally sunny weather? When the spring tides are over—"these big tides"—the weather will break and we can expect a few days of rain. How long will this gale last? Wait till the tide starts ebbing; it will level off then. In other words, during the day a marked change in the weather can be expected to coincide with a change in tide. Over longer periods, weather trends will mysteriously follow the variations from neap to spring tides and back again.

3

In any island fishing community, life is concentrated on the shoreline. Man becomes a littoral animal. The shore—almost entirely that of the harbors and coves—is the islanders' contact with the sea, which is their means of transportation and commerce with the mainland, their chief place of work, the source of their living, their wealth, their happiness, and their sorrows and disappointments. Road frontage means little, but shore frontage is cause for higher tax assessment than that of property lacking access to the water. A house with land fronting on good herring grounds in the harbor, though served by a rutted rockpile of a road, may be taxed twice as high as a similar house facing the paved highway but having no shore rights. The ideal of every waterman is to have a gentle-sloping, sandy beach of his own where he can keep his punt, pile his traps, and haul up his boat for repairs and wintering, or perhaps build a wharf. The fisherman without his own shore rights is somewhat like a farmer who doesn't own his land. The man of importance in the community is apt to be one who can say, "That's my beach," or "That's my wharf," or better yet, "I own that frontage and rent privileges to three or four of my neighbors."

From earliest childhood the islander's life has been oriented to the waterfront. Aside from the food he gets from it in clams and mussels and the fertilizer in the form of rockweed, it is his nursery and playground during the first fifteen years of his life. Here the island boy or girl grows up. Here every child, for a time at least, becomes nature's playmate and pupil. In the pools he spies and captures in toy traps the tiny lobsters and crabs. As he herds the schools of minnows to his play weirs, he learns that innumerable individual beings at times act as a single organism. Under the rocks that appear at low tide he finds perfect starfish, and maimed ones quickly adjusting to a life with four points. As the tide returns he is aware of one of nature's basic rhythms, and as the meadow of seaweed floats upright in the rising water lifted by its chain of buoyant balloons, and the barnacles and sea urchins unflex their silky feelers in the cool new water, he sees the resourcefulness of nature in equipping all her creations for survival. The larger fish—flounders and pollack—that come in with the heightening water give him inklings of the reservoirs of life in the depths beyond low water. There are, further, more definite inklings when a fisherman ties up at the neighborhood wharf with several tons of cod, hake, and haddock, and perhaps some curiosity such as an overgrown sculpin, a blue shark, a monkfish, a sea cucumber, a huge skate.

No less fascinating is the mammalian life of the island waters. Around the ledges, especially those a little offshore, the seals dive and swim. During the day they flop on the rocks and their barking carries inshore. Sometimes an otter or a muskrat swims across the head of a cove. And on boat trips to and from the island an escort of porpoises often appears, playing tag around the boat to the general delight of young and old. The waterman who can be blasé about a porpoise does not exist. More rarely a whale wallows up to the surface, much larger than a workboat, awe- and fear-inspiring to all who see it.

4

Overhead, both alongshore and on the water, the air sparkles and flashes with gulls and terns. Black shags or cormorants, sea pigeons, and eiders skim the waves. Most numerous, most involved in the life of the islands, is the multitudinous herring gull. Wheeling overhead, riding the waves among the moorings, perched on the spilings of every wharf and on the ridgepole of every fish house, this bird is always present, and the only time he is not graceful is when he is waddling about on the beaches with his close friend, the fish crow, with whom he is always ready to share a dinner of garbage. The air from dawn to sunset is resonant with the gull's wild cry—the "barbaric yawp" that Whitman sought to emulate.

The herring gull is a worker and an entertainer who supplies both comedy and poetry. On any wharf where the men gather to gossip and to loaf, they will always have half an eye for the sea birds. The greedy gull choking on a fish twice too big for his throat causes an amused chuckle. Two gulls in a tug-o'-war over a herring bring guffaws. But in the air his gracefulness is the wonder of ornithologist and fisherman alike. Over the harbor, over the bay, and far out on the horizon above the trawlers, the gull traces his gliding circles against the sky. He rests on the air currents or sails against them, without flap of wing. When he descends, he drifts down in languorous spirals or plummets in a nose dive, sometimes rocking from side to side. He skims along the wave crests, extracting fish from the water while in full flight, without wetting a feather. He drops splashlessly upon the surface, feet thrust forward and wings upspread like a diver's arms.

No shore dweller questions the usefulness of the gull, which is the great scavenger, the cleanser of the sea. Anything remotely edible thrown on the shore will disappear. Even though there is no gull in sight, before a person has gone fifty steps from where he has emptied the bucket, the first bird will glide in. At first he will fly in tilted circles, the lower periphery curving to within a few feet of the water where the garbage floats. Then others soon appear and finally the whole flock settles on the water for the feast. Behind every lobster boat, every trawler returning from deep water, follows a cloud of gulls to snatch up scraps of old bait and entrails tossed overboard as the fishermen dress their catch. The gulls perform their greatest service in the harbors around the fish wharves and canneries. The livability of the community depends on them. In their absence the stench would exceed human capacity to suffer.

At one time the fishermen, particularly the islanders who lived near the breeding grounds, plundered the gull. The eggs are highly palatable; the fledglings, when taken home and fattened in a coop, provide excellent meat. In early summer, picnics and Sunday outings used to be organized to rob the nests. Laborers in the quarries among the islands would bolster their income, or even attempt to earn a living during unemployment, by egg collecting. A pail of fresh water was the candle; fresh eggs sank to the bottom, incubating ones floated. When the gulls were thus driven from the inhabited islands to the outer ledges and islets, fishermen on their way back from "outside" would stop off for a bit of nest robbing. Later, when milliners paid handsomely for the breast feathers of gulls, the species faced extinction.

Though at present the gulls live in close partnership with man, they are wary of him. They no longer breed to any extent on the mainland of this country. Excepting Criehaven, on whose rugged southwest headlands is a vast nesting colony, they prefer not to share an island with human beings. About all they will tolerate is a lighthouse keeper, as on Great Duck Island and Matinicus Rock. A populous breeding place is an awesome sight. Overhead thousands of birds rotate and spiral in a clamorous, scintillating cloud. Among the blueberry bushes, on the coarse brown grass, under the scrub spruces, among the driftwood and rocks alongshore, and on the naked ledges are the nests with their three brown-speckled eggs, or with one egg unhatched, another hatched, and still another with only a tiny beak chipping at an already partly cut peephole on life. Some nests appear deserted, until one looks in the grass or among the rocks and finds the brown-speckled young, sometimes only a few hours old, where their instincts have told them to hide.

The gulls prefer to roost, too, far from human dwellings. Many of the offshore ledges are snowy with excrement. On some of the islands that they frequent, their feces have killed the trees. Grassy islands have benefited from this rich fertilizer so that many sheep can be grazed on mere islets.

5

Back from the shore Swans Island is a waifish fragment of the Maine wilderness. Of its nearly 6000 acres only a small portion, perhaps 1000 acres, was ever cleared, and today at least half of that has slipped back into forest. The remaining fields are small and, like the ocean around the island, are broken by jagged ledges. Everywhere the greedy spruces and alders are devouring the cleared land. In only a few places is the soil deep and level for farming. These places are mainly along the northeast shore and in the interior of the "north" as the section beyond the Carrying Place (actually to the west) is called. Generally the terrain is hilly, rocky, or marshy.

In the eastern and larger segment of the island are several hills. One of these, Goose Pond Mountain, soars to the Himalayan altitude of 240 feet. On the ledges near its peak is a pile of boulders which some islanders, tongue in cheek, maintain are the ballast of Noah's Ark. "How else could those boulders get way up there?" Big Mountain, just to the west, attains a height of 210 feet. The rounded summits of the "mountains" are bare, fissured granite. The spruces on them are stunted; only bayberries, blueberries, and lichen flourish there.

In the middle of the island is Goose Pond, covering four acres between the two highest hills in as wild a setting as any watering place of moose and deer in the northern forests. On a small island a freshwater pond has a great distinction. It is a rarity, something unique in all the vastness of surrounding salt water. The possibility of fresh-water fish, which most have never tasted, is alluring. When the state stocked the pond with trout, some of the islanders got out their cod lines, like small rope, and filed the points of hooks that were an inch across. Arguments developed as to whether clams or salted herring made the best trout bait and whether the usual two and a half pound sinker was necessary. An expedition was organized to build a raft on the lake, and many fathoms of rope and lobster warp were donated to secure it to the shore when not in use. Finally, after consultation with sporting acquaintances "ashore," several islanders sent for proper gear and caught some trout, but the catches have never been outstanding. Hunters have better luck at Goose Pond. Black duck breed there and many alight on it during the autumn migrations. Deer are plentiful here at the center of the island.

On the shore of the pond one is as far from the sea as he can get. Older inhabitants have nostalgic memories of sleigh rides there in winter and bathing parties in the summer to enjoy the unwonted warmth of the water. Except for the gulls that come in great numbers to splash in the fresh water and wash the salt from their feathers, the ocean may be forgotten here. Frogs, pitcher plants, water lilies, dragon flies, bloodsuckers, all the plant, insect, and animal life of mainland ponds is in abundance.

The natural life of the island is astonishingly varied. Red and gray squirrels and chipmunks chatter in the trees. The meadows and swamps are full of hares, the descendants of two pair introduced many years ago. But undesirables like skunks, woodchucks, and poisonous snakes either never lived on the island or have been exterminated. The man who trapped the last fox in the arctic winter of 1918 is regarded as a public benefactor. Garden snakes and toads are numerous. Most of the common birds of the woods and fields exist in greater numbers than on the mainland. A few islanders, aware of the remarkable resources of their home, are ardent and successful bird watchers. Most of the trees, ferns, and wildflowers indigenous to Maine grow on this island, often—as with wild roses and iris—more luxuriously because of the milder and damper climate, and fewer people to molest them.

6

The climate of the Maine islands differs greatly from that even a few miles inland. Winter and summer extremes of temperature are much less severe. On Swans Island readings below zero are rare. Whole winters have passed without the ice on the ponds getting thick enough to cut. The all-time low is —20°, but that may have been the vagary of someone's faulty thermometer. Heavy snows do occur and with them there is generally higher wind than away from the coast. Yet the snowstorms of the mainland are often rain or sleet on the islands. A storm often begins as snow but changes to freezing rain which forms a thick, drift-proof crust. Swans Island has a snow plow, but there are winters when it is seldom used.

The winds are raw and violent. When it blows from the

northwest it brings the coldest temperatures. Vapor rises from the bay and forms a thick coating of snow-like frost in the grass along the shore. One's face burns as with dry ice, and no combination of woolens and oilskins is protection against the icy claws of the gale. The receding tides leave a girdle of ice around the shore, and sometimes the coves freeze sufficiently for skating and ice-boating.

The southerly winds are warmer, but they build up rougher seas. So unflagging are the winds, no matter what their direction, that weeks may go by without the lobstermen getting to their traps, which are in deep water at this time of year. Only Old Harbor offers safe mooring for a boat. Those who in other seasons fish out of the coves either move to the harbor or drag their boats onto the land until spring. But whether their boats are drawn up or not, this is the time the men work ashore. Some still cut next year's fuel supply in their woodlots. Occasionally the women accompany them to enjoy the calm of the winter woods where the chickadees chirp and flutter among the spruce boughs, far from the wind-buffeted savagery of the shore. Others in their workshops repair and make new lobster traps for the coming spring and summer. The men are around home much of the time. Neighbors visit frequently. The social life of the island is renewed.

In late March and April the winds begin to moderate. The returning plankton dyes the cold blue water of the bay with tints of green. The boats are returned to their summer anchorages. But damp raw gusts still rattle out of the north, and southerly winds carry little promise of relief. May passes, a seasonal disappointment. Then, in the first two weeks of June, spring crowded by summer sweeps over the island; but from the land to the north, where it has crept down to the coast, not from the south where the ocean currents are not much warmer than in midwinter. White and purple lilacs bloom on bushes as big as trees. Apple, wild plum, and cherry blossoms burst open, and the timid green of the hardwoods brightens among the conifers. Insects and birds fill the air with sound and motion. Fires finally burn out in the heating stoves, and for the first time since September people appear on the roads without woolen shirts and jackets. The spring fogs are over. A tide of limpid blue air drifts out of the northwest, etching the shores and inland hills in as deep a blue and clear an outline as on the sparkling days of midwinter. The sails of pleasure boats glitter on the bay and glide above the spruces as they pass between the islands.

Midsummer brings more haze and in August the heavy fogs march in off the ocean like multitudinous invading armies. They overrun the south shores first, then stream up the channels among the islands and spill out along the northern shores, till the land is oppressed under smothering layers of dampness. For days, sometimes a whole week, the fog whistle at Duck Island mourns through the dripping atmosphere. During the daytime the fog may lift a little from the interior of the island, and sometimes it will withdraw from the bay entirely. But on the ocean side it will cling tenaciously, so that a walk of a mile overland will carry one from thick to clear weather. In the evening the mists slink back to stifle the island in opaque blackness all night.

In July and August the temperature sometimes mounts into the high seventies. Reports of heat waves come from the mainland, and on the island the mercury may top eighty, but only on a day or so and for a few hours till the saving southwester breezes up. At these times the islanders speak darkly of heat prostration and curtail their daily tasks. Summer people remark how pleasant this is, really, compared with what they are accustomed to at home. The natives pretend great wonderment and display their discomfort all the more. But the "heat waves" seldom last. Either the fog intervenes with its delicious chill—a liberating army now whose advance guards are hailed joyfully as they probe up behind the islands—or the wind shifts to northwest and drenches the bay in white-capped coolness.

Autumn is the island's season. The fogs creep in more rarely. Long successions of northwest days with clear gentle air set in. On the north shore, at a spot where the road skirts close to the water, the people stop their cars or pause on their bicycles to admire the view up the bay to Blue Hill and the villages sparkling like whitecaps at its base. At night the sky is sometimes alight with the aurora. And as the fisherman rows his punt in from the mooring, the wake and the water roiled by the oars sparkle with sea fire. This is the time of the big herring runs. Lobstering still continues, but evenings and moonlight nights the fishermen are busy with twine and torches to net the fish so necessary for bait and so valuable at the sardine factories. Ashore, from the middle of September on, the ground at dawn may be whitened with frost. The leaves turn and are wrenched from their boughs by the wind. But there are not enough hardwoods to provide lavish displays of color—only a patch of yellow or red here and there among the spruces. The fields brighten more as the blueberries turn crimson and purple. The woods and marshes become festive as the leaves fall and reveal the scarlet clusters of white alder which grows here as nowhere else. In late October or early November the first snowflakes drift down from iron skies and the lobstermen think of changing their moorings and of the winter's work indoors and in the woods.

2

Settlement

IN THE EIGHTEENTH CENTURY ON BRITISH ADMIRALTY CHARTS Swans Island was named first Brule-Cote and later Burnt Coat Island. Many local antiquarians must have pondered the fact that this whole coast was first mapped by one of the greatest of all navigators, Samuel de Champlain. An explanation for the name was devised, its basis being a faint knowledge of French and a total ignorance of Champlain's descriptions of his voyages.

Ask any islander for the origin of the name of Burnt Coat Harbor. He will paraphrase for you Dr. H. W. Small's account as given in his *History of Swan's Island*. This book, written by an island physician almost sixty years ago, your informant will describe as "the real history of the island and you can find it in the library in Augusta." Dr. Small's words have a confident ring—so much so that their substance has appeared in numerous books on the Maine Coast, accepted unhesitatingly as authentic and not as legendary. Dr. Small himself had no misgivings. The account was older than he; its seed must have been planted long ago at the fireside of some sea captain musing over his charts and letting his fancy play over stories he had heard of the explorer who had first sailed these waters.

Dr. Small writes: "The first European who visited this island is not known. The first authentic record was made by Champlain during his voyage along this coast in 1604. He made a map of the whole coast and gave names to many islands on either side of us. . . . Champlain gave the name of this island on that early map as Brule-cote [sic], 'brule' meaning burnt, and 'cote' hill—Burnt-hill. It is supposed that Champlain designated the island by some hill that had been burnt over. Some later discoverer translated 'brule' burnt, but did not translate 'cote,' hence on his map he incorrectly gave the island the name Burnt Cote. Another more stupid still, thought the former had made a mistake in spelling, and on *his* map had Burn Cote, by which name it is called in a deed given October 28, 1790. . . . Later it was generally known as Burnt Coat or Burn Coal Island.

"It is quite probable that Champlain visited and explored this island, as would seem likely by the accurate map he drew of this and the neighboring islands."

A variant of this theory is that Champlain had "Brun Côté" on his map, meaning "Brown Side." The trouble with both versions is that no map attributable to the explorer contains these names. In this area Champlain named only Mt. Desert Island and Isle au Haut, two mountainous islands that would serve as unmistakable landmarks for future navigations. He did not clutter his map with names of what he describes as "numerous islands, rocks, shoals, banks, and breakers on all sides, so that it is marvelous to behold." The space between Mt. Desert and Isle au Haut, where Swans Island is situated, he filled with a sprinkling of ink dots, all of the same size and shape, apparently to discourage any later voyager from attempting a passage that way. In his travels in this region, Champlain had gone aground and damaged his boat more than once. "For greater security," he wrote, "one should keep the sounding lead in hand."

Other explorers than Champlain sailed the Maine coast. John Smith, Verrazano, John Cabot, even Lief Ericson must have seen the Mt. Desert Hills rising above the many-islanded sea, and the broad bay of the Penobscot cutting deep into the northern wilderness, even though they did not attempt to map them. From the Gaspé to the tropics are no such landmarks as these, and it would be a sorry navigator indeed who failed to notice them and, so doing, failed to head shoreward to observe them more closely. Anyone thus deviating from his course would skirt along the barrier of islands, of which Swans is one of the largest and most prominent, searching for a spot to put in for water or for an entrance into the bay waters beyond. Surely one vessel from this early parade of adventurers veered briefly out of line to drop anchor in Burnt Coat Harbor.

In all probability, European fishermen frequented the island at an early date. Basques, Normans, Bretons, and Portuguese thronged the Newfoundland Banks during the century after Columbus' discovery. By 1600 they had drifted westward to the New England coast. Arriving for a summer's fishing, they would pick out some harbor where they could anchor safely, find adequate stocks of fuel and water, and dry their fish for the voyage home. Sometimes they pastured domestic animals on small islands nearby, where they might be left from year to year to multiply and fatten—a permanent supply of fresh meat and milk.

The most persuasive evidence that Swans Island served as a fishing station is its harbor. It is storm-proof, secluded, yet easy to find because of its proximity to those peerless landfalls, the Mt. Desert Hills. Above all it is accessible to waters swarming with fish. Mariners bold enough to sail three thousand miles in vessels not much larger than a modern sardine seiner could hardly fail to discover and make use of it. But evidence is not conclusive. The fishermen did their best to keep their fishing grounds to themselves. Whatever charts they had, they did not circulate. Their buildings and encampments ashore were too flimsy to last out the centuries.

2

There is an aboriginal, an Indian, era in the island's history, one whose evidences are unmistakable and plentiful. Scattered about the shores are shellheaps, some of them still apparent on the surface, others covered with a foot or so of soil. The shells are mainly those of clams, though some of oysters and sea urchins have been found. Mixed among the shells are bones of fish, waterfowl, and deer. Island tradition has it that these heaps are the remnants of oft-repeated orgies of feasting. The Indians were the originators of the "clambake," a form of repast still highly regarded in New England village life.

Numerous Indian burying places have been uncovered on the island. Just below Hocomock Head is a tongue of land where the plows of early settlers turned up so many skeletons that the spot bears the name Burying Point. The skeletons, which were found just under the sod, were cf a people of larger frame than were the Indians known in the region in later times. One wonders if perhaps these were not the bones of European fishermen who died there, but there is no way of telling now. Other burying grounds were at the Carrying Place and at a place called Middle Head. Only a few years ago a summer person, temporarily archeologist, discovered a human skull, presumably Indian, on tiny Phinney Island just north of the main island.

Hocomock Head is the only place on the island that retains its original Indian name. Scholars translate the word *hocomock*, which appears elsewhere as a place name in New England, in two ways. Some say it means the "evil spirit." Others say it means the "place on the other side." Indians who were encamped on Harbor Island, one of their favorite sites, would refer to this promontory across a narrow channel as *hocomock*, "over there."

Before the Indian wars of 1725–26 the tribe that frequented this region was the Malecites. These were salt-water Indians, expert with the canoe and skillful hunters of seal and porpoises, their chief sources of food. Indians of this tribe may have lived more or less permanently on Swans Island. The burial places are very likely theirs, and many of the skeletons may date back to the great plague that in 1618–19 destroyed three-quarters of the native population of the Maine coast.

From the mid-eighteenth century almost to the present, the Indian visitors to the island were Penobscots, with headquarters at Oldtown, though some of their close friends, the Passamaquoddies, came also. The Penobscots had always had an outlet to the sea by way of the river which bears their name. But they were never a tide-water tribe. Their excursions among the islands were strictly summer outings. About June 1, after they had planted their corn, beans, and potatoes, and the black flies and mosquitoes had become unbearable in the inland forests, they would drop down river in their long, low-riding canoes with the tribal brown and white triangular decorations painted just below the midsections of the gunwales. The first night they would camp beside the river near Bucksport. The next morning they would fan out into the open water. Birch bark and ash paddles would now be pitted against the chop and current of the bay, sometimes against the ocean itself, as hunting parties paddled down beyond Owl's Head to the seal grounds at Matinicus, fifteen or twenty miles offshore. Present-day lobstermen still wag their heads in incredulous admiration when they speak of the skill and daring of these canoe-men.

Swans was one of the most distant islands that the Penobscots visited. Possibly they did not arrive there until the third day after their departure from Oldtown, though the route down Eggemoggin Reach is direct and sheltered. Once encamped they would set about their summer's business, for this was an important phase of the tribal economy. From their canoes they hunted seal and porpoises for the skins and oil so necessary in their cooking. Sometimes, like the Malecites, they would eat the flesh of these mammals, but they did not relish it.

While the men were engaged in this dangerous occupation on the water, the women would be robbing the sea fowls' nests of eggs and young birds, which they considered great delicacies. Or they would gather sweet-grass for their basket work, or dig great quantities of clams and search beneath the rockweed at low tide for lobsters, both of which they would dry and pack in bark containers for winter stews and soups. The shellheaps are perhaps not so much the results of on-the-spot feasts as of the accumulation of years of clam drying.

The Indians continued these visits to the island to within the present generation, though they came less and less frequently as the tribe declined in numbers and adopted white men's ways. On Swans Island there are no records or traditions of trouble between the races, though tales of murder and revenge are told of Indians and early whites on other islands. The first English settler on Matinicus lost his life after Indians discovered the buried bodies of several fellow tribesmen he had murdered. But that was before the Revolution. Swans Island was settled after the Penobscots had joined the Americans against the British. In 1779, despite Mohawk threats, the most famous of all Penobscot chiefs, Joseph Orono, pledged his tribe to the Colonists' cause and thus, historians claim, "saved Maine to the United States." Every child on Swans Island knows of Orono and is proud that, according to tradition, this island was his favorite summering place. To commemorate his visits one of the islets off the north shore has been named after him.

3

The first white settler in what is now the township of Swans Island was a fugitive from the law and from life. Thomas Kench had come to the colonies from England before the Revolution. When war broke out he enlisted for the march to Quebec. He labored up the Kennebec with Arnold in late September, the batteaux slicing the red-gold film of leaves drifting downstream in the autumn sunlight. He portaged through the wilderness to Dead River and camped beneath Mt. Bigelow, already capped with snow as a warning of what was to come. After torrential rains he fought upstream against the freshets, passed the seething junction with Alder Stream, and staggered beneath the boats on the portages between the tiny mountain ponds that formed the precipitous staircase to the height of land. He watched mutinous companions give up and lurch back down the trail to starvation; he saw men fall in exhaustion to die in the frozen woods. He dined with the others on moosehide moccasins or on the hindquarters of a dog. And after days of wallowing in the swamps around Lake Megantic, he was present on the upper reaches of the Chaudière when the men fell like wolves on the live cattle sent up by foragers from the French settlements. Later he tented on the Plains of Abraham during November and December while smallpox flamed through the camp. He was in the attack on the Citadel, was one of the first to gain the walls, and one of the few to escape from them alive. Finally he was among those who outlasted the lingering Canadian winter and the return trip through the sodden forests to New England.

In 1776 Thomas Kench joined the artillery company of Captain James Swan, who was later to become proprietor of Burnt Coat Island and give it his name. Apparently after signing a "mutinous" petition to his superiors, Kench deserted from the Army and sought peace, alone, among the outer islands of the Mt. Desert region. On Harbor Island he found the seclusion he was seeking. He dug there a cellar and built a log house, cleared a piece of land, and brought on a cow and some sheep. In the harbor he planted an oyster bed, apparently with some success, because many oyster shells have been found near the site of his dwelling.

For fourteen years Kench lived there with his disgust, his hate, his fear—the first in a line of island solitaries extending down to the present day. The nearest white settlement was ten miles away across land and water on Mt. Desert. Fishermen would sometimes put in at the harbor, as they had been doing for generations, but Kench avoided their company. To them he became a figure of legend, a recluse living unnaturally like a medieval hermit, or a sort of Cyclops skulking on his rocky island with his sheep and his cow. The Indians, too, during their clambakes broke his seclusion from time to time, but the savageries of the redmen must have seemed rather tame when compared with those he had known on the Quebec expedition. And the shock must have been rude when in 1786 his old captain, James Swan, brought boatloads of laborers into the harbor to build saw mills and fell the forests.

Yet these years of island life did much to heal Kench's spiritual hurt. In 1791 the New Hampshireman David Smith settled with his wife and children almost on top of him on Harbor Island, and Kench didn't move away. Smith had been a soldier in the Revolution; he had fought at Bennington, been wounded at Bemis Heights near Saratoga, had been honorably discharged and presented with a lifelong pension. In 1786 he had come to nearby Deer Isle in the wave of emigration from the older states into the unsettled portions of northern and eastern New England. Here he had married the daughter of a settler from Gloucester, Eunice Thurston, by whom he had sixteen children. Five years after his marriage, the Smith family landed at Harbor Island, presumably with four or five children. Kench's solitude was broken, his privacy a thing of the past. Yet he remained for five years to watch the Smith brood increase to two figures. As a matter of fact, he did not depart till after the Smiths had left to take up residence on the main island. Perhaps the silence was too oppressive for him then, or perhaps another neighbor, Joseph Prince, the agent of the new owner of the islands, was not to his liking. At any rate in 1796, twenty years after his flight there, Thomas Kench was cured of his addiction to loneliness and moved to Brooksville on the mainland to live out the remainder of his more than ninety years. He bought a hundred acres, became a farmer, and quickly married, not too late to father six children.

David Smith was not only the first permanent settler on Swans Island but also one of the chief contributors to its population. A true patriarch, he was known in his lifetime as "King David," and is still affectionately referred to by that title. It would be difficult to find an islander today in whose veins some of his blood does not flow. During his eighty years he had three wives. Of the first, whom he married in New Hampshire and by whom he had three children, little is known. Neither she nor her children came to Maine. Of the sixteen children of his second marriage, eight lived beyond childhood and several married on the island. After his second wife's death Smith, not much past fifty, married again and sired eight children, most of whom lived to perpetuate the line. He owned a farm of 240 acres "at the North," where he produced bumper crops and kept his herds to feed and clothe the multitudinous brood which he housed first in a log cabin and later in a frame house forty feet square.

The organized settling of Swans Island was unusual in its methods for so late a period in New England history. Commonly a town was established by a group of persons petitioning the provincial or state government for the right to settle in a specified unappropriated area. The petition was usually granted, with certain stipulations as to the establishment of schools and churches. Later the inhabitants would request incorporation into a plantation or town. The founding and growth of such a community is described by Mary Ellen Chase in *A Goodly Heritage*. But sometimes, and especially among the islands, a settlement was started as a proprietorship, a business venture for the benefit of the entrepreneur. Such a venturer was Colonel James Swan of Boston.

Swan had come to Boston in 1765 as a boy of eleven from Fifeshire, Scotland. He worked in several counting houses and made many friends, among them Henry Knox, later Washington's Secretary of War. At the age of seventeen he wrote an impassioned book entitled "A Dissuasion of Great Britain from the Slave Trade," which makes him one of the earliest abolitionists. As an ardent member of the Sons of Liberty, he took part in the Boston Tea Party. When war began he fought at Bunker Hill, where he was twice wounded, and later he served as an artillery captain in the siege of Boston. Before the end of the war he held the posts of Secretary of the Massachusetts Board of War and Adjutant-General, and had been elected to the General Court. He was owner or part owner of several privateers and further advanced his fortunes by buying and selling confiscated Tory property. Among his friends he numbered Lafayette, Rochambeau, and Washington. After the war he continued to hold a prominent place in politics, business, and society. He speculated wildly and lived habitually beyond his means. He boasted of owning the most ornate carriage in North America. In his role as a great gentleman, he fought a duel at twenty feet. His was a colorful, reckless nature, bound eventually to destroy itself.

In 1786 Swan obtained from the State of Massachusetts a deed for Burnt Coat Island "and all the islands, the centers thereof are within three miles of the said Burnt Coat Island." There were twenty-five islands in all, aggregating 9623 acres, for which Swan paid in Commonwealth securities whose face value was £18,000. He and his family referred to their new holdings as their "Island Empire." From it they hoped for rich profits and much pleasure while residing there summers and entertaining their friends in baronial style.

The islands abounded in timber, much of it hardwood, which could find a seller's market in the new communities along the coast where home-building, ship-building, and lime-burning all demanded vast cordages of wood. Many men were capitalizing on these opportunities created by the post-war expansion of New England. Before the Revolution Royal Governor Bernard had attempted, and failed in, a venture similar to Swan's when he had become proprietor of a large part of Mt. Desert Island. But things were different now. Around 1790 General Knox developed extensive holdings at Thomaston. A little later the huge Bingham grants in eastern Maine were the cause of frenzied speculation in which both Knox and Swan figured prominently.

The state exempted Swan's islands from taxation for twenty years if within seven he induced twenty-two Protestant families to settle there in houses at least twenty-two feet square, and if he constructed a church and a building in which to hold grammar school. But Swan's purpose was not to build churches, schools, and durable homes for his settlers. They were left to build their own cabins out of logs and to plaster the chinks with mortar made of crushed clam shells from the heaps left by the Indians.

Swan first built a mill to saw his lumber and grind his workmen's flour. The mill, like many others on the Maine coast at this time, was operated by the tides. There are no brooks on the island with enough water to saw heavy timber. To reduce island forests to planks, the tides had to be harnessed. As one islander puts it, "Those old fellows had to use what they had at hand,"—a guiding principle of all New England life in its early days.

On the east shore of Burnt Coat Harbor is a cove of about eighteen acres, with a little island midway in its entrance. At low tide this cove is empty with the exception of a foot or so of water in a small area near its mouth. At high tide it is filled to a depth of eleven or twelve feet. Swan had two dams constructed of logs and rock, one on either side of the islet, blocking off the passages between it and the shores. In the center of each dam was a floodgate and millrace. The sawmill, which was situated to the south of the islet, could be operated only on the ebb. The grist mill, on the north side, could be used on both the ebb and the flood. Both mills could not be run at the same time.

Swan himself very likely supervised these installations. Among the many facets of his brilliant but unscrupulous mind was a strong mechanical bent. For example, on his first trip to France he was outraged to find eight men at work beating flax where a machine designed by him on the spot could do the same work and save the expense of wages. France, he concluded, was industrially backward because it had too much manpower.

On October 26, 1786, he was able to write to his friend Knox: "Returned last night from the islands—I found them astonishingly prosperous—every shilling expended yields great returns." He expects to make £600 annually from two or four saws. His advances of money have been considerable, but "profits are in proportion." His tone is jubilant, light-hearted. "I was drove down in the wind—we flew—at the rate of thirteen knots an hour—I was not wanted in the other world, or they would not have let so good an opportunity have slipped."

To get his island enterprise started Swan had had to borrow heavily, and by the next year his finances were in collapse.

One saw was all he ever operated, and his dreams of great profits proved empty. The mills have been in disuse now for over a hundred years. Yet the rock embankments are still there, and at low tide the barnacled timbers emerge briefly.

The millstones have had a more prolonged life. After many years of grinding island corn and barley, they were taken to Brooklin, the nearest point on the mainland, to be used in a mill there. Later a Swans Island blacksmith and gunsmith, Jacob Reed, brought them back and set them up in a mill he had built on a brook on the eastern shore, and there once again they ground the islanders' flour. In the meanwhile a vein of "blue rock" that traverses the northeast tip of the island was found to contain traces of gold. A gold-mining expert was brought from the West. "Pretty looking" stocks were printed and sold, but not to islanders, who "weren't that gullible." Steam engines and stonecrushers were bought and assembled. For the final pulverizing, Swan's old millstones were harnessed to steam. A shaft was sunk, the ore was carted by oxen to the crushing sheds built near a brook which supplied water for sluicing. The hole in the hillside grew deeper and the pile of slag higher. After several years of labor the men and oxen and steam engines had produced enough gold to make one good-sized wedding ring. Operations were suspended. The shaft filled with water, and after a stray cow had drowned in it, it was filled up. The millstones for a time lay covered in the rubble near the engine house. Then one day an islander who had just built himself a huge house with a bulbous mansard roof dug up the stones and placed them in his front yard to serve as a flag-pole standard. There they remain today, relics of Swan's "Island Empire," which in fact turned out to be as abortive as the gold mining of later years.

At the same time that he was installing his mills Swan was building himself a manor house on the slope just north of the mill pond, where there was a view out past Harbor Island to the open sea. The big colonnaded mansion, which was completed with the speed of all Swan's enterprises, served as model for Montpelier, the residence that General Knox soon after built in Thomaston, where in emulation of Swan he was developing his own private principality. Montpelier, reconstructed, is one of the show places of Maine; it is hard to envisage a similar edifice on the scrawny, granite shores of Minturn. In true feudal style the settlers called Swan's establishment the "Big House," but it never fulfilled its promise. It soon became a sort of island tenement, with as many as nine families living in it at one time. King David, moving over from Harbor Island around 1795, took squatter's possession of it as his temporary palace. In 1840 the house, already in a state of collapse, burned to the ground. The islanders took the bricks from the foundations for use in their chimneys. The cellar hole is long since filled—a square, grassless depression in the jungle of an alder-choked pasture.

Because he needed settlers and laborers to fell his trees, to raft the trunks around to the mills or haul them overland by oxen, and to saw them into boards and timbers, Swan announced that he would allot one hundred acres, to be deeded over free of charge at the end of seven years, to anyone willing to bring his family to the island, build a house, and clear a piece of land. He had no trouble finding takers. But after a time he modified his offer. Through his resident agent, Joseph Prince, he agreed to convey ten-acre plots to any fishermen who would settle on the island, live there with their families and livestock, pay their taxes, and maintain such roads as Prince deemed necessary. As loyal villeins they were to have their grain ground in Swan's mill and their lumber cut by his saw and shipped in his vessels. Many of his settlers were soldiers who had fought in the Revolution against this sort of paternalism on a national scale. One begins to understand why Swan's old gunner, Kench, decided to move to the mainland after his twenty years at Harbor Island.

From its inception the toy empire had many troubles. For one thing Swan did nothing about the school and the church. Many of the first settlers did not remain. Most of those who came to stay did not arrive till after 1800, perhaps because Swan and his agent made the islands too much like a remnant of George III's empire. Nor were the transactions that Swan entered into only five years after his purchase encouraging to prospective settlers. On October 28, 1790, while he was in Paris, his Boston lawyers sold the whole Burnt Coat group to his agent for £300. Prince continued Swan's policies and honored his agreements with the settlers. Then in 1795, Prince sold the islands, again for £300, to one of the attorneys who had acted for Swan in the previous sale. A year later this attorney, Henry Jackson, sold the islands back again to Swan for the same sum.

By these manipulations, which were typical of his methods, Swan must have hoped to keep the islands out of the hands of his many creditors. He was peculiarly badly hit by the postwar depression. By 1788 he had to flee incognito to France and was writing of entering a monastery to learn French so that he could begin his career anew in that country.

When the Revolution broke out he became the agent of the French government in purchasing American goods to be smuggled through the British blockade. In 1795 and 1796 he was in this country in that capacity, operating with great secrecy. His major achievement was the settlement of the American debt to France, whereby the French gained much needed dollars for their American purchases. But in 1796 the Comité de Salut Public, accusing him of malfeasance, stripped him of his agency. Soon he was back in Europe again fleeing his American creditors. During this frenzied sojourn in the United States he had no time to think of his islands; he had left them in charge of his agent, Joseph Prince, who had full power of attorney.

Prince eventually departed to the mainland. The deeds promised at the end of seven years were never delivered to the settlers who had lived up to their part of the agreement. The settlers commenced to appropriate any land that looked good to them. Squatters' rights were the only ones they recognized. After his years of absence they probably considered Swan was out of the picture. And they had grounds for optimism. In 1808 Swan was jailed for debt in the prison of Ste. Pélagie in Paris. Protesting that the claim against him was unjust, he refused to pay and remained in jail for twenty-two years. When he was released by revolutionary decree in 1830, he died shortly after in Paris.

Yet even from jail Swan continued to bandy the islands about on paper. In 1812 he mortgaged twelve or fifteen of them, including the home island, to a Baltimore friend and creditor named O'Maley, who received the mortgage so that other creditors could not get the property. Five years later O'Maley foreclosed and engaged Daniel Webster to look after his interests. Swan, acting in close concert with O'Maley, also hired a lawyer to protect his interests there; the two attorneys, in fact, were to act jointly in Swan's and O'Maley's names. Later these attorneys turned the account over to a Maine lawyer. His unpleasant task was to collect money from the settlers for the property which they had occupied for many years and which, in some cases at least, Swan had promised they could have for nothing. The lawyer went to the islands, assigned deeds to each settler, and then took a mortgage on the property to secure payment for it. The islanders ignored the mortgages and the attorney attempted no action.

By 1834, Swans Island had sufficiently disentangled itself from the proprietorship to seek some form of self-government. The population had increased to almost two hundred.

Without the customary petition to the legislature, the citizens met in a private home, organized the modified form of government known as a plantation, and elected officers. In the same year the most pressing community needs were attended to. The tax list was drawn up and a road constructed. At long last a school house was built and in it classes were held for two and a half months in the winter and three months in summer. A small but significant action in the first plantation meeting was a vote to release a certain islander from financial obligation to O'Maley.

Up to this time the island had been a pioneer settlement on the eastern frontier. From now on it was an established community eligible for state and county aid in road-building, support of the poor, and education.

3

The People

BY THE 1950 CENSUS THE TOWN OF SWANS ISLAND NUMBERS 468 in population, the composition of which is indicated by the following table prepared by the Bureau of Census.

	Male	Female
Total	227	241
Native white	219	241
Foreign-born white	8	—
Negro	—	—
Other races	—	—
Under 5 years	28	38
5 to 14 years	31	39
15 to 20 years	13	16
21 to 24 years	11	14
25 to 34 years	34	36
35 to 44 years	29	18
45 to 54 years	23	31
55 to 64 years	27	24
65 years and over	31	25

The population of the island declined steadily from 765 in 1880 to 452 in 1940, the normal trend of back-country New England in which maximum populations were almost invariably reached in the nineteenth century. The slight increase since 1940 is the result of the war and post-war birthrate, for in 1950 there were sixty-six children under five as against thirty in 1940.

At one time there were four clusters of population on the island, but with the almost complete desertion of "the North," which has no good anchorages, there are now only three—Swans Island village (locally called Old Harbor), Minturn, and Atlantic. Minturn and Swans Island are situated on the shores of Burnt Coat Harbor; and Atlantic, somewhat more spread out, occupies the northeastern side of the island, where there are several summer anchorages and where most of the cultivable land is situated. Each of these localities, among which the population is about evenly apportioned, is a tiny community in itself with its post office, church, stores, wharves, and, until recently, school.

One can search the voting lists of Swans Island for "foreign" names almost in vain. There is a family with a French name that stretches back to Acadian times. There are two Swedish families who came years ago to work in the granite quarries and remained to become fishermen. They are now so much a part of the town that they are no longer thought of as aliens.

Another type found on the islands from time to time is the homeless outlander, often a Scandinavian, who is left behind by some work gang or appears on the shore seemingly from nowhere, and who lingers on for years or even a lifetime. Living in a shanty or a tumbled-down house on the "end roads," he merges into the island life and, though existing apart in linguistic and social isolation, serves some obscure psychological need in the community—the need perhaps to have someone to protect and to feel superior to.

The residents of Swans Island are almost all descendants of families who were living there one hundred and fifty years ago. These first settlers came from the mainland and nearby islands, where in most cases they or their parents had recently come from some eastern Massachusetts fishing village like

Gloucester or Marshfield, the home of their forebears since the early seventeenth century. Pressure of population and insufficient opportunity to make a living had for some decades been forcing families to sail eastward down the coast in their Chebacco boats in search of a cove or inlet in which to settle. If the fishing was poor or the land sterile in one spot, they would soon move on. So long as they had their boat they could seek a new home. When they found what they wanted, they settled, as permanent as the rocks that lined the shores.

Perhaps Swan's offer of land and work was the main attraction for some families, though not for all. These people were fishermen, not laborers. Swan himself recognized this when he later attempted to entice fishermen who he hoped would bring business to his mills, help fill the population quota set by the Commonwealth, and perhaps work for him in the winter. But most of his lumbering was done by more transient workers. At one time he employed a gang of Irishmen, who as soon as the trees were felled vanished, leaving behind the name Irish Point given to the locality where they camped. Of their labors only one relic remains—a heap of cobblestones that they dumped into a gully over which they hauled their lumber.

Most of the settlers whose descendants are still on the island came after 1795, at a time when Swan's control had already been reduced to a paper one only. Racially they were Anglo-Saxon, and today Swans Island perpetuates its English origin as undiluted as any village in America or even in England itself. Not even the other British nationalities—Welsh, Irish, Scottish—have gained a toehold there.

At present most of the residents bear one of eight family names, all of which have been current on the island for well over a hundred years. Two or three of these names are so numerous that they account for half the population. There is much intermarrying. Family relationships become so complex that any degree of kinship less than that of aunt, uncle, nephew, niece, and first cousin is ignored. An island girl marrying an island man has a good chance of retaining her own name. There are cases of women who have kept their own name through two marriages. Marriage of first cousins occurs, but not often. It is frowned upon, but not as much as elsewhere.

The person looking for a Juke or Kallikak situation will be disappointed. No "effects" from inbreeding on the minds and bodies of the residents are noticeable, unless it be a slightly higher average intelligence and more robust physique than are generally found on the mainland, and these superiorities probably result from the vigorous and varied island life. Most of those who leave the islands to live and work elsewhere are successful above the average in their new environments.

The men sometimes find their wives in families off the island, but unless they go away from the coast the chances are that they will marry into their own kind of stock. This marrying almost exclusively close to home is understandable today when Maine fishermen seldom sail out of sight of their harbors. But it was equally true in the last century when the men roamed the world. The tradition has been for the man to seek his wife in native territory.

But on all the islands there is at least one limited source of new blood—the women who come to teach school. Because salaries are small in these tiny communities, the teachers who come to them are usually on their first job. They are young and sometimes pretty. But above all they have the attraction of being new, something different among the perhaps too familiar island faces. Also, the teacher may often be homesick, eager for company, and hence friendly. Thus many an up-country farmer's daughter has ended as the wife of some island fisherman, and certainly the island has been the better for it. Similarly, the daughters of physicians and ministers who have worked on the island have occasionally remained as housewives.

2

The islanders share many characteristics in common. Most outstanding, and most often celebrated in novels of the Maine coast, is a highly developed sense of independence. Dorothy Canfield Fisher in a recent book wrote that Vermonters "hold these truths to be self-evident: that life is not worth living unless every individual can get air enough to breathe deeply and freely. But nobody can help him do his breathing. It is an intrusion to try." Maine islanders are even more convinced of these "truths" and act upon them more consistently than does the Vermonter. This individualism is, of course, a part of the older New England tradition. It flourished during the time when the island was a frontier settlement and did service to the islanders during the latter part of the Swan regime, when their rights were being so outrageously flouted. But Swan and O'Maley were ousted. Eastern Maine ceased to be a frontier, though many of the communities retained the frontier state of mind. On the islands, which were physically and mentally out of the main stream of national development, this lag was exaggerated.

But extreme independence may be harmful to oneself and to others. The first literary enthusiast of Maine coast life, Sarah Orne Jewett, was impressed with the individualism among the people of Tenant's Harbor whom she depicted, with a transparent fictional covering, in *The Country of the Pointed Firs*. Yet for all the keen observation in this book, which still is the classic of the Maine coast, Miss Jewett is just sufficiently sentimental not to see that highly developed independence as it appears in some of her characters is at best an anachronism and at worst sheer folly. As a realistic reporter of New England life in her time, Miss Jewett portrays several village recluses. But she can only praise the "independence" that leads persons to live apart in a world of psychotic dreams or physically to withdraw onto some windswept outer island where winter and summer they subsist on clams and defy the town "fathers" who plead with them to return to human society. Mary Wilkins Freeman, a contemporary of Miss Jewett, was able in her stories of New England solitaries to grope through to the tragedy, the spiritual death, that this psychopathic individualism could lead to.

Recluses like those in the pages of Miss Jewett and Mrs. Freeman are no longer as common as formerly in the mainland villages of New England, where at one time they constituted a tradition and had a world-famous representative in Emily Dickinson. On the islands, though, the tradition has not faded. Many men and women live alone in houses once designed for a family of ten or twelve. Some are not unsociable or unuseful. They mingle in the community and do their quota of work. Others are shy, withdrawn, real spiritual shut-ins who will step into the woods to avoid nodding to you when you meet them on the road. In their solitude they all develop a psychiatrist's catalogue of personality distortions —miserliness, misanthropy, and obsessive fondness for animals, usually cats, which they keep by the dozen. And, as in Miss Jewett's sketches, the community not only leaves them unmolested in their cocoons of isolation but even applauds their self-sufficiency in managing to exist for a few dollars a year on a diet of salt-fish and pilot biscuit.

As a community, the islanders glory in being removed from certain mainland influences. They are most law-abiding in serious matters, but they rather pride themselves on disregarding the law in petty activities like hunting and driving. "No one will bother you on here," the man who drives an unregistered car will say. Shooting out of season and jacking deer are common practices. From time to time the youth have spells of breaking into closed cottages either out of deviltry or in search of liquor. When complaints are lodged, the local authorities resort first to persuasion or appeals to fair play to set things right. "How would you like it if someone broke the windows in your house, or almost ran down your child?" Arrests would do little good. These are almost family affairs, much better settled on the island without dragging in people "from away." Thus civil cases also are seldom taken to court. The law enforcement officers act as umpires, unofficial magistrates or squires, to settle these disputes as to rights-of-way or access to a well. Persuasion, reason, compromise are their weapons, and usually they are effective weapons.

Where this independence, whether of the community or the individual, is self-sufficiency and not just fear of the outer world, it deserves applause. In fact, it is necessitated by the way of life of the islanders. On the coast and in the back country certain callings have to be followed in solitude. The trapper, the forester, the farmer plowing the far fields had better be amused with their own company; otherwise they will be most pitiably bored. Some fishermen and most lobstermen do much of their work alone; they become early and deeply schooled in loneliness. A generation ago a lobsterman often put in a whole summer by himself on some scrap of an island where he could keep his traps, pull up his peapod, and build himself a rickety shelter. That was one of the accepted ways of lobstering and it was profitable. A man would have the island's shores to fish exclusively for himself. The price he paid was solitude broken only by the weekly visits to the mainland to sell his catch and stock up on provisions. But that type of lobstering is no longer done. When the motorboat made it possible to set traps at a greater distance from home, a man was no longer able to pre-empt an island all to himself simply by camping there. But here and there a shack once occupied by such a man still remains teetering on the shore of some treeless outer island.

Charles William Eliot in his *Life of John Gilley* gives an account of the inner resourcefulness of the older island men. William Gilley, father of John, went at the age of sixty-three, when the youngest of his ten children had grown up, to live on Great Duck Island—a near neighbor of Swans—where he remained until he was nearly eighty. His immediate purpose in going to this desolate offshore island was to raise sheep. But isn't one permitted the luxury here of believing that this islander secluded himself like some Biblical figure with his flocks because of a need to assess the accumulated resources of a lifetime? His withdrawal was into a far different domain than the deserts of sterility in which Miss Jewett's neurotics lived and still do live here and there among the islands.

Persons like John Gilley have not entirely disappeared from the Maine islands. On Swans alone could be found several older persons whose lives have achieved a fulness approaching his. A notable example is a woman of eighty, a widow and the mother of two sons and two daughters. Her formal education was limited to the grades. Her girlhood and young womanhood on one of the island farms were years of hard work. Yet she was not crushed by circumstances and as advancing years brought relief from drudgery, she decided to see what could be made of life. Living now in the upstairs rooms of the family farmhouse, she cares for all her own needs, even to cutting her own firewood. With no professional aid she has taught herself to play the violin, organ, Spanish and Hawaiian guitars; she subscribes to *Etude*, which she reads to the last period. To keep her mind agile she has studied French and Latin from the textbooks used by her grandchildren in various stages of high school and college. Among the French books she has read is Pierre Loti's *Pêcheurs d'Island*, a copy of which one of her sons found in a dump on the mainland and sent to her. She found the book interesting but the ending, in which a young fisherman is drowned at sea, leaving a pregnant wife behind him, depressed her. The same situation occurred in the life of her own mother.

Annually she reads through the Bible in its entirety. She is a champion of the King James version, questioning the authenticity of the newer translations. Theological matters of all kinds occupy her mind as they have occupied few minds in the past hundred years. No problem is too petty and none is too momentous to escape her cogitations. The difficulties in applying the Golden Rule give her much food for thought, but never doubt. She has observed that at times when she has treated people as she would like them to treat her, they have become angry with her. Of course, she attends church regularly, helps with the Sunday School, and plays the organ when asked. She is a relic of the New England of several generations ago, in which education and religion were the most vital matters in life and in which one was expected to develop his personality himself in whatever direction he saw fit.

3

The spirit of independence shows itself in other ways. One is fanatical protection of what the islanders consider to be their personal rights. Any invasion of his privacy the islander will combat savagely, whether the invader be a fellow islander or an outsider. Protracted and dreary feuds develop over the location of a fence or the use of a shore privilege. Embarrassed constables and the deputy sheriff will be called in. Occasionally an extremely litigious person may journey to the county seat to consult with lawyers and examine deeds. Many novelists have dealt with these situations—notably Ruth Moore in such books as *The Weir* and *The Fire Balloon.*

If an outsider such as a minister, a physician, or a summer resident attempts some change on the island, say the clean-up of a dump, the organization of a Scout troop, or the improvement of boat service, each person asks: "How will that affect my rights?" The populace is most wary when it is offered something for nothing, a wariness that might have had its origin in Swan's empty promises a hundred and fifty years ago.

Being vigilant, cantankerously so at times, the islanders have been able to avoid many undesirable developments. No disproportionate summer colony has grown up to reduce them to menials. No large business has flourished to the decay of individual enterprise. The island is a community of self-employed men and women—more so today than ever before. Town individualism has fended off many a half-baked project common on the mainland.

But personal individualism has stifled many a worthwhile project. Until recently the islanders have shied away from cooperation; the kind of cooperation that in pioneer days led the whole community to join in a barn-raising, or in constructing a wharf. Cooperation meant surrendering some minute part of one's theoretical rights. A man's barn totters to the ground because he can't afford to hire labor to mend it and his neighbors won't join to help him. A wharf used by a half dozen fishermen slips off into the water, yielding a foot or so of its length in every winter storm because the men will not organize to repair it. "Everyone is waiting for the other fellow to fix it," someone explains, "and in the meanwhile it's tumbling into the drink." The island has been left for years without an adequate passenger boat because the townspeople couldn't agree on what port she should sail from.

There are signs that the trend is changing. After much agony, a recent island physician managed to get an electric-power cooperative started under the R.E.A., and in the building of a new school some of the men gave time and labor free of charge. But many other vital projects requiring community action remain undone.

Though they need to learn how to cooperate, the islanders are always ready to help a neighbor in trouble. The minister of one of the churches recently received a purse of $400 when he had to have a major operation. His own parish consisted of only a few dozen persons, but members of other churches and some who belong to no church at all contributed. They liked the man. He had done many odd jobs to eke out his ministerial wage and he had always been an obliging and reliable worker. But the main reason he got help is that it's an island tradition to lend a needed hand. House fires bring out the whole town. If a man's boat goes aground or breaks its moorings, all he need do is inform the telephone operator; in ten minutes half the men of the island will be on hand.

There is no lack of friendliness here. More neighborly people do not exist, so long as there is no infringement of real or imagined rights. Once an islander gives his friendship, he withdraws it just as reluctantly as he would withdraw his enmity. All islanders are ready to chat in a friendly way with any acquaintance—and sometimes interminably. One doesn't know what a welcome can be till he has experienced one in an island home after he has been away for some time. The older men and women frequently address one as "dear," no matter what one's sex. First names are invariably used. Family histories are divulged in detail at the slightest provocation. One becomes wary of asking favors of island friends because he knows they will disregard inconvenience and expense to please one. You happen to mention that you would like to go handlining. Though you forget your remark the next day, you will soon hear that a half-day's expedition is being organized, involving the use of a motorboat and fishing equipment belonging to the islander. At the end of the trip is the refusal to accept a cent, even in payment for the gasoline.

Yet the outsider is never on the same footing as the islander. His position is not necessarily unfavorable; it may even be more comfortable; but it is different. You are either "from the island" or "from away." If you are a stranger, a visitor, you will find that being "from away" is no disadvantage. You will be treated well; the hospitality due a stranger will be accorded you. You will find listeners for what you have to say, will receive much attention, and feel at home in the stores, in the churches, and on the street. If you come as a summer resident your status will be more complicated. But in general you will receive the same treatment as the more casual visitor, so long as you make no pretense of not being "from away." The fact that you own your summer home and are on the tax rolls is of little importance. Compared with the islanders with their one hundred and fifty year family tradition on the island, you are "from away" and will remain so always.

If you come to practice some profession, as a teacher, a physician, or a minister, and if you do your job faithfully and well, you will be tentatively accepted. Your origin off the island will be almost, but not entirely, overlooked. But if you come to the island to go into business, be prepared for uphill work. It will make little difference if you come from Los Angeles or from some little hamlet across the bay. You are from off the island. If you are married to an islander, it will be a help, but not so much as you might expect. Until your flesh and bones blend with the earth in one of the island cemeteries, and until long after, people think of you as being "from away."

Occasionally a businessman or artisan smashes through the barrier. But it's more difficult than on the mainland where people are used to hiring someone from the next village or buying something there if they can't find the man for their job or the article they need in their own community. Loyalty to old connections lingers on the island. Island business should be given to islanders as much as possible. It's bad, of course, to have anyone go broke, because that will add to the town's relief expenses. But if someone is to go broke, it is better that he be a newcomer. He may leave; but the islander will remain, an embarrassment to the finances and pride of the town.

4

To go with their way of life the Maine coastal islands have developed a distinctive manner of speech which is partly unintelligible to the outsider. A conversation between a summer person on Swans Island for the first time and a native fisherman is apt to consist largely of, "What did you say?" and "I beg your pardon?" And the visitor need not be from a thousand miles away. He could have almost as much trouble if he was born and raised in Bangor seventy miles to the north. As a matter of fact, the severest critics of the Maine coastal dialect are the citizens of Maine from a few miles inland. They fail to see anything forceful or attractive in the speech of these people whom they sometimes refer to as "herring snatchers."

The speech is a variation of general New England pronunciation, the characteristics of which seem always to become exaggerated near the salt water. On the peninsulas and islands the exaggeration is extreme. To describe this dialect is impossible; all attempts by novelists to reproduce it have been failures; but a few of its more salient features can be given. As in all of eastern New England, *r*'s are dropped at the end of syllables unless they are followed directly by a vowel in the next syllable or word. "Come heah, deah. Here is youah milk." In certain cases an "intrusive" *r* is inserted between a syllable ending in a vowel and one beginning in a vowel. "Soar (saw) up this wood." This sounds like straight New England speech, but on the islands there would be a difference in accenting. Words like *dear, here* definitely take on an extra syllable in the form of an accented *ah,* not a neutral *uh* or *eh,* as in most of New England. The interjection "Oh, dear!" is thus pronounced "Oh dee-ah!"

The vowel *a* in words like *past* and *answer* is broad as in British. But the *a* in *ar* coming in the middle of words like *hard* and *garden* carries to an extreme the general New England tendency to flatten—to become the *a* in *sad* rather than the *a* in *father.* Thus *Bar Harbor* becomes *Baa* (like a sheep's bleat) *Habbah,* again with all syllables equally accented.

The sound of e in *get* or *ie* in *friend* becomes *i* to make *git* and *frind,* as is so often the case in back-country speech. The sound of a in *many* and *any* and similar combinations becomes *i* also, to make *minny* and *inny.* The most remarkable feature of Maine island dialect is the treatment of y at the end of a word—a feature that Celia Thaxter noticed on the Isles of Shoals nearly a hundred years ago. The y in such words as *Benny* becomes ay or perhaps the French é. Again, the final syllable in such words receives a much heavier stress than in the speech of other localities. If a fisherman meets you coming from the shore with a clam rake and basket he is apt to say, "Gittinnay?" which means, "Get any?" The two words are run together and accented so as to rhyme with and correspond in meter with the French word *Dubonet.*

All these characteristics are exemplified in the following sentence: "Binnay, come ovah heah. Soar up this hadwood stump faw may." The tempo, which cannot be shown by any trick of orthography or diacritics, would be rather fast and very even, though the last syllable of a sentence or phrase tends to end in a lilt or accent. The combining of words into sentences and phrases of which the rhythm suggests that of a single long word is quite reminiscent of French. The sentence, also, is more musical than ordinary American speech because fewer vowels are slurred in unaccented syllables. Anglophiles among the summer people like to think that the islander has retained much of the British in his speech and in proof they point to the treatment of *r*'s and the use of the broad *a*. A more subtle resemblance is in these speech rhythms. The Englishman and the islander would ask the question "Are you there?" quite similarly: "Ah you theah?" What would make them sound alike, even more than the dropping of their *r*'s, would be the lift at the end—the refusal to slight a final syllable, even if it is one they have added to a monosyllable.

There is great variety of speech among individual islanders. One dialectical hazard, especially among men, is excessive speed. Some have so accelerated their sentence tempos that the effect to unfamiliar ears is that of a rapidly spoken foreign language. Syllables and words slide together into an oriental sing-song. Another hazard, also confined to men, is a stentorian loudness. A few men bellow even in close conversation as if they were trying to make themselves heard over a quarter of a mile of water in a hurricane. Perhaps it is the need of rapid, loud speech in fishing and boating, when decisions have to be made, conveyed to one's fellows, and carried out instantaneously, that has made men more susceptible to these habits than women. The women, at any rate, speak more slowly and softly—in short, more restfully.

Those who feel that geographical variations in language are an asset to a nation but fear that the radio, movies, and TV are going to smother them in a universal colorlessness may discard their worries so far as the Maine islands are concerned. In the children and young people the regional pronunciation is more marked than in their parents, and in the parents more than in the grandparents. Yet the two younger generations have been exposed to the drab articulations of radio announcers all their lives. The threat of uniformity is not great where one hears a little girl say to a playmate: "Jimmay, hold youah hosses. What's the hurray?"

Sarah Orne Jewett, carried away by her enthusiasm, spoke of the Chaucerian quality of the Maine coast vocabulary. One would have to strain one's imagination unduly to hear the voice of Chaucer, but with careful listening one will catch an occasional faint echo from the age of Elizabeth I. As one steps into a tippy punt the owner may say, "Watch it! That punt's quick!"—quick in the sense of alive, as in the Biblical "the quick and the dead." Or a woodsman may tell how he spent his day "clefting," *i.e.*, splitting, logs. And the remark, "I had a terrible hoist (tumble) on the rocks," recalls the Shakespearean "Hoisted on his own petard."

Phrases which are regional or dialectical rather than archaic are more common. Rather than say a wind is rising, the islander would phrase it, "An air is breezing up." If a fisherman has had a battle with foul weather crossing to the mainland, he'll say he had a "hard chance going off today." But, he may add, he "toughed it out and gained the day." If he were a real old-timer and an electrical storm had added to his "hard chance," he'd say that the "thunder squabble" gave him quite a time. At least it wasn't an "out wind," that is, from off the ocean and thus very strong, though he was afraid it might "haul off to the south'ad and come thick o' fog." But the rain was "something sinful."

As he told his story you'd notice the almost Victorian quality of his expletives. The fashion for hard swearing has kept free of the islands, though on certain occasions the islanders can let go with an artistry second to none. But in ordinary talk their expletives are either namby-pamby ones like "Goodness gracious," "Heavens above," and "Oh, dear" or equally innocuous but slightly more distinctive ones, like

"For the love and honor of God," "By chowdah," "My soul and body," and "By the gods o' war."

The application of nautical terms to daily life is, of course, common and is exploited in every story and description of Maine coastal life, so that it need not be hashed over again at length. But one or two from Swans Island seem somewhat original. A man suffering from rheumatism may say, "I guess I'll lay to till I get my back unsnarled," the imagery being that of a fishing boat suspending operations till it gets its trawl lines disentangled. Or a woman after a very cold night will ask you if you "kept fire watches all night," that is, got up every three or four hours, the period of a watch on shipboard, to stoke the stoves.

Certain other terms which seem highly localized are worth mentioning. "On here" means "on the islands," and carries connotations of the special status of islands, their independence of mainland ways and responsibilities. It is usually spoken with a slight smugness. "On here we don't bother with game laws," or "On here we can't afford to pay teachers three thousand dollars a year." A "concert" is usually not a musical affair but a pageant, or play, or program of any sort. A "social" is any gathering where the audience-spectator division doesn't obtain but everybody participates.

Finally island talk is rich in simile and metaphor. "Hard as a half-tide ledge"—the ledge that, visible only from half to low tide, mercilessly tears out the bottom of a boat—is used to describe a piece of oak or a woman's heart. If a person chokes at doing something, is disgusted, or highly unwilling, he is said to have "swallowed a bone." A highly conscientious or strict person is "straight up and down," usually uttered with an accompanying vertical movement of the hand. A first child who ceases to be the center of interest when a second child is born is described as "having his nose broke." Of a man of peculiarly vigorous constitution it may be said, "He was so tough that his heart didn't stop beating till after he died."

The Maine islander is pleasant to listen to once you become accustomed to his "accent." His voice never has the metallic qualities of the uneducated speaker of the eastern cities nor the nasality of back-country New England. His grammar is amazingly pure. *Ain'ts* and double negatives are the exception. Most of his errors, by schoolmarm standards, are in the use of archaic or dialectical forms of participles—*broke* for *broken*, *blowed* for *blown*, *borned* for *born*. His vocabulary, particularly the figurative part of it, is extensive because he has two funds of words to draw from—that of the sea and that of the land. He enjoys his own talk as much as do his listeners "from away." Unlike many of his Yankee cousins inland, he is a ready and tireless talker. With no people in America, not excluding the Middle Westerners, is it easier to strike up a conversation. The islanders have the ease and pleasure in talking that marks people who are at ease with themselves, their environment, and their neighbors. Their speech, vivid, pure, original, is one of their finest attributes, a witness to the soundness of their way of life.

4

The Fishing Industry

RECORDED ON THE LITTERED SHORE OF BURNT COAT HARBOR IS the story of Swans Island fishing, which is basic to an understanding of the people and their way of life. In miniature it is also the story of much of maritime New England. Every foot of waterfront contains some relic of past activity: forgotten and uncountable enterprises in menhaden, lobsters, codfish, mackerel, and herring.

On the west bank, in Swans Island village, the buildings of a fish-packing plant, from which salt cod used to be shipped to Russia and the West Indies, have been partitioned off into compartments where lobstermen store their bait and gear; and the surrounding wharf gapes with holes that threaten broken legs and bone-crushing plunges to jagged rocks below. Farther up the harbor is the abandoned steamboat wharf, its spilings loosened and its fill sliding into the water. Near it are the foundations of a sardine factory that once employed a hundred men, women, and children.

Tossed up on the east bank, in the village of Minturn across the harbor, is the boiler of an old cod-liver-oil factory. Next to it the buildings and wharf of another salting plant are slipping into the water. Nearby are the timbers of a lobster cannery. At the mouth of a little cove, elongated rock heaps perpetuate the memory of Swan's mills, the first island enterprise. And down near the entrance of the harbor is the great quarry wharf, the terminus of the little cable railway that once lowered blocks of stone from the gouged hill above. The wharf, built of mammoth blocks to defy time itself, now serves as a landing for the present mail and passenger boat. Back from the shores are other derelicts—a schoolhouse with glassless windows staring imbecilely; the tottering behemoth of the Redmen's Hall; the wreckage of machinery and boilers strewn around the abandoned quarry. With all these reminders of bygone eras the islander in his speculative moods, which are frequent, lives in the past tense. In pointing out landmarks and spots of interest, which he loves to do, his formula is not: "That is . . ." but: "That used to be . . . the cannery, a boatyard, a sheep farm."

Yet amongst all these ruins and shards, so depressing to the sightseer who has been conditioned to posters of neat, white-spired Maine fishing villages rife with flower gardens and picket fences, are other enterprises humming with unprecedented prosperity. Lobster buyers' wharves with their floats and scales and gasoline pumps and stores for selling paint, hardware, and gear, and the fishermen's sheds festooned with rainbows of lobster buoys give evidence of an enduring vitality.

Writers of the coast of Maine, from Sarah Orne Jewett onward, have bemoaned the rotting wharves and silt-filled harbors. When the prosperity of a village has entirely vanished the laments are justifiable. But more often they are not. Romanticists frequently confuse signs of change with signs of degeneration. The way of life of a community may have altered radically, say, from the China trade to lobstering, or from codfishing off the Banks to boarding summer visitors. But the change may not have been for the worse. The community may be enjoying the greatest prosperity, the most varied and the richest cultural life in its history. All this wreckage of the past that so troubles writers points to the inexhaustibility of the offerings of the sea and the versatility of the people themselves. So many changes have been lived through, so many crises have been survived, that one becomes confident that the next crisis, even be it the disappearance of the sorely beset lobster, will not find the island without resource.

In the earliest years of the island's settlement, fishing was only one of many occupations in a way of life that was largely self-sustaining. Each man had sufficient cleared ground to raise all the hay, vegetables, and grain his family and animals would use. He could pasture sheep anywhere—often on some islet where fences were not needed—and these supplied him with meat and with wool for the women to card, spin, and weave into clothing. In the fall he shot ducks and geese and in the spring he robbed the gulls' nests. In the woods during winter he cut fuel and logs to be sawed into building lumber. For ready cash as well as for his own needs, he fished, usually with handlines, though for herring he would use dipnets. Since fish were plentiful, he could count on a surplus to sell either fresh or dried. The salting of fish and drying it in the sun on "flakes" was part of the housework of every island woman. Even to this day some of it is done.

If all else failed the early settler, he could turn to the shellfish on the mudflats and nearby bottoms, still an important source of food and income for the taking. Down to the present mussels are gathered occasionally when the canneries demand them, and in the reaches north of the island a few boats do some winter dredging for scallops with reasonable success. And the clam has always been most important as bait and as a free palatable food. Without clams many an early New England settlement would not have survived its first winter. Contemporary accounts of Plymouth Plantation contain grateful praise of these shellfish which the settlers believed were placed there by God to preserve His saints. Early comers to Swans Island also found the clam a staunch friend in the lean first years. Island families have always appreciated this rich, tasty food at their doorsteps.

2

Until about fifty years ago the fishermen of the small Maine towns and islands had several very prolific deepwater species to choose from so that they could concentrate on catching the one that at the time was bringing the best returns or was most abundant. They were not the victims of the natural vagaries of any one fish and its market. If one species disappeared or could not be sold, they caught some other.

The first extensive commercial fishing was for cod. As the island began to emerge from a subsistence economy shortly after 1812, the islanders began taking voyages along the coast as far as the Gulf of St. Lawrence in search of bigger catches. A stimulus came from the passage of a Congressional act in 1819, which provided for bounties to the owners of fishing vessels employed in cod fishing. From the first, the bounty laws required that a benefiting ship engage in cod fishing for at least four months a year, though not necessarily all in one stretch. In the final law the bounties were fixed at $3.50 a ton for vessels between five and thirty tons burthen and four dollars a ton for vessels over thirty tons. The master and three-fourths of the crew must be American citizens and the fishermen must be paid by a division of the catch or a share in the price brought from its sale. Only the cook could be paid in regular wages. These laws not only encouraged the fisheries but kept in readiness a corps of skilled native mariners for service in case of war.

The requirement of four months spent in cod fishing annually made longer voyages desirable and longer voyages necessitated larger boats to carry supplies and bring back the larger catches. Also the schedule of subsidies favored boats of greater tonnage. Hitherto the most widely used fishing craft along the coast was the Chebacco boat, named after the Massachusetts parish in which it was first constructed. This sturdy craft, which had brought many of the early settlers down the coast of Maine, ranged from five to twenty tons in capacity and was high-stemmed, sharp-sterned, and only partly decked. In the bow was a small cabin with bunks and a cookstove. The lines, particularly the sides, of the Chebacco were suggestive of a barrel. But now a larger boat, the pinky, was developed. Though it was essentially a Chebacco on a bigger scale, its rounded stern was lifted much higher and the cabin was more spacious, but still dark and stuffy. Roomier and more seaworthy, it was suitable for trips as far as the coast of Labrador.

Finally schooners came into use, proving themselves yet more serviceable for the Bank fishing that the Swans Islanders engaged in for about fifteen years. The first of these schooners was *The Constitution*, which in 1853 made a voyage to the Grand Banks. Soon three or four others were also visiting the Banks each summer. On their return in the fall they entered the mackerel fishery, which was becoming the dominant island occupation.

Life on these voyages, whether in pinkies or schooners, was rugged enough, though it had its compensations. On the fishing grounds the hours were long and the work exhausting. The baiting, setting, hauling, or underrunning of trawls, the cleaning and curing of the catch, not to mention the sailing of the vessel, made for round-the-clock working. Yet discipline was extremely easy. These were crews of equals sharing in the profits of the venture and thus working for themselves as much as for the owners, though the owners, who paid the outfitting bills, naturally got the largest share. The Maine boats were manned almost exclusively by men of native stock, all of the same background. Many members of a crew would be blood relations. The master had no need to be like the tyrants who commanded the motley crews of the whaling vessels. Certainly there was no need for the regimentation found on a man-of-war. Often the captain would consult with the men before making decisions. The food was necessarily simple and the quarters confined. Rum, particularly in earlier times, was a staple on all voyages. Sometimes a keg was lashed on deck, a cup handy, so that the crew could help themselves at will. In the hold beneath the salt required for curing the catch other barrels of rum would be stowed, partly so that it wouldn't be drunk up at once and partly as an incentive for the men to work hard using up the salt.

3

The Bank fisheries gradually became the monopoly of the big schooners from ports nearer the consuming centers. Island fishermen lacked the means and the desire to compete. Finally in 1866, the bounty laws were repealed. No Swans Island schooners visited the Banks after 1868. But even while cod fishing was at its height, mackerel fishing was gaining all along the coast.

For years mackerel had been caught exclusively by hook and line from Chebacco boats and pinkies. In the early days the men would fish while under full sail because mackerel were thought to take only rapidly moving bait. Later the technique of tolling developed. The men scattered chopped bait, "chum"—herring, menhaden, or clams were used—to attract the fish, which were then caught from a slowly drifting boat. Soon bait mills were invented for grinding the chum, which formerly had been minced laboriously with hatchets. With tolling came a new type of hook, the jig, whose shank was imbedded in lead to make it more manageable in the water. When the vessel entered a school of mackerel, the toll bait would be spread and the crew, sometimes as many as twenty strong, each with four to six lines, would stand along the rail and hook the fish as long as the chum, which was constantly being sprinkled in the water, lured them.

Each man slatted his fish into his own barrels because he was paid in accordance with what he caught. Sometimes a good man could fill as many as ten or fifteen barrels in a day. A boy could do this kind of fishing as well as a man, and probably get more fun out of it. Lads of ten and upwards were customarily among the crews of these schooners. Mackerel "jigging" was truly a family affair, and it must have been the realization of many a boy's dream when he first shipped on a cruise with his father and brothers, perhaps for the whole season. Beginning so young, a youth sometimes commanded his own vessel before he was twenty.

About 1870 the hook and line gave way to the purse seine in mackerel fishing, and the Swans Islanders, skilled in the use of net in catching herring, came into their own. The bigger boats would start their season in April off Virginia and Maryland. As the summer progressed and wave after wave of this mysterious fish beat in on the coast, the schooners worked northward where they were joined by the rest of the fleet. They ended the year's work in late September or October in the Gulf of St. Lawrence or the Bay of Fundy. They almost always landed their catches at Boston, Gloucester, or Portland. In 1879 the Swans Island fleet landed almost fifteen thousand barrels of mackerel at five dollars apiece. After the mackerel season was finished, the schooners would try for a fare or two of cod or perhaps have a go at the West Indian or coastal trade.

Swans Islanders were the most skillful mackerel fishermen on the coast, unmatched even in Gloucester. For fifteen or twenty years their energies were poured enthusiastically into this activity and with phenomenal success. Their skippers were sought to command boats from other ports and fishermen from other towns coveted berths on the highly successful Swans Island boats. In 1881 twelve vessels from the island with crews totaling 180 men were on the mackerel grounds. During the fishing season it must have been a woman's island, for almost all the male population was seining somewhere along the coast. Probably no other New England community had so large a proportion of its productive workers in one so specialized fishery. Portland and Boothbay, both much larger, had only thirty-five and thirteen mackerel boats respectively. But more impressive was the high rank these ten or twelve Swans Island boats took among the New England fleet, which numbered almost three hundred. In 1881 three of the ten most successful boats were from here—the schooners *A. E. Herrick*, *Isaac Rich*, and *Alice*.

The pre-eminence of the Swans Island mackerel fleet is vividly exemplified by the schooner *Alice*, commanded by Captain H. B. Joyce. Year after year the *Alice* ranked first or among the high-liners in its fares. In 1880 she cured 3,700 barrels of fish worth $19,548; in 1881, 4,905 barrels worth $28,055. She often was the first to make catches in the southern waters in the spring, one year netting thirty thousand fish in a single May day off Block Island. She was among the few ships ever to make a sizable haul with purse seines in the Gulf of St. Lawrence, which for some reason could ordinarily be fished effectively only with hook and line. So successful was her skipper that he had a steamer specially made for mackerel fishing—supposedly the largest fishing steamer in the world at that time. He continued to make large catches with this ship, which carried a crew of forty. Yet the high cost of operation and a sudden falling off in the supply of mackerel in 1886 forced him to sell her at a loss.

The disappearance of the mackerel from the Atlantic coast

was sudden and almost final. During the previous twelve years the catch for the whole New England fleet had averaged one hundred million pounds. In 1885 about one hundred and twenty-five million pounds were landed. In 1886 the catch was only thirty million pounds, and in no year since has the total climbed much above that figure. The reasons for this catastrophic decline are not clear. Probably man-made causes had little to do with it. The explanation lies hidden in the ecology of the ocean and in the instincts of the mackerel which annually migrates shoreward from deep water in search of food and spawning grounds. These inexplicable reversals in the long-established habits of a species are among the many hazards of the fishing industry. In the 1890's the mackerel fleets of Swans Island and scores of other communities were gradually dismantled—the ships sold or converted to other uses.

This was not the first such trick that fish nature had played on the islands. During the 1870's, just before the mackerel boom, the whole Maine coast went on a frantic spree of catching menhaden, or porgies, to convert into oil and fertilizer. The menhaden, which is a species of herring, frequents southern waters in vast schools. For nearly twenty years during and right after the Civil War it invaded the Gulf of Maine. At Blue Hill a "porgy press" was invented to squeeze out their oil. In the '70's factories were erected all along the coast to handle the enormous catches. On Swans Island fishermen invested in vessels, nets, buildings, and wharves for catching and processing the porgy. In 1875 two oil and guano factories were built at Old Harbor, one of them furnished with steam boilers and hydraulic presses. In 1876 Maine manufactured almost three times as much oil from this fish as all other states combined. Then in 1879 the menhaden failed to appear; henceforth they spawned and fed elsewhere. That year not a single oil factory opened and not one man was employed in the industry in the entire state. The porgy has never returned to eastern Maine in usable numbers. The factories on Swans Island had operated only a fraction of three years.

Many islanders had invested heavily in this ephemeral industry. Some did not recover their losses. Yet it was by no means a general calamity because the mackerel fishery was already thriving and awaited great expansion. It soon reached boom proportions, and the magic wand of prosperity touched the island, bringing about undreamed-of transformations. New and pretentious houses were built; horses replaced oxen which had hitherto been used almost exclusively; new roads were constructed and a steamboat began making scheduled stops to lure the inhabitants ashore to spend their newly acquired money. Fishermen, like all those who live by luck, are free spenders. They are examples of New England thrift only when they have little to be thrifty with. These were the island's "good old days," still lamented by the older inhabitants, when its population reached its greatest number—almost eight hundred—and there were two hundred and fifty boats. Today the population is about four hundred and fifty and the number of workboats about ninety.

4

But the vanishing of the mackerel and menhaden did not mean the end for Swans Island fishing. Lobstering had already got started, and still plentiful among the islands was the herring, which from earliest times had been netted for bait and for the family table. Some fishermen had built smokehouses and cured a surplus which they could sell or trade on the mainland. Later an increased demand for smoked, pickled, and canned herring—not to mention its continued use as bait in all kinds of fishing—made the herring of major importance in eastern Maine. By the 1880's Swans Island had nine vessels of over twenty tons in the herring fleet—a number exceeded among New England ports only by Deer Isle in the Bay of Fundy and by Eastport. For every herring boat of over twenty tons there were a dozen smaller ones. The skill that they developed in this kind of fishing is what gave the Swans Islanders their great advantage in mackerel seining—made them, in fact, the best net fishermen of the Atlantic coast.

The simplest and one of the oldest ways of catching herring is by torching. A flame is burned on a pole projecting from the bow of a small boat at night. If there are herring present, they will be drawn to the surface by the light, and a man standing in the bow with a dipnet can scoop great quantities into the boat. This method isn't profitable in commercial fishing, but a few men still employ it for catching bait. It is certainly a picturesque way of fishing and is consequently a great favorite among novelists who purport to describe Maine coast life.

Another old and tried way of catching herring is by the brush weir—pronounced like *wear*. Introduced from Nova Scotia into the Eastport area as early as 1820, the weir soon became a familiar sight east of Portland. It fishes particularly well on a broken, rocky coast. On Swans Island and adjacent islets, there are half a dozen or so active weirs.

A weir is an enclosure made by driving piles into the mud during ebb tide and interweaving them with brush. Leading up to the enclosure are two sweeping barriers of brush, one of them starting almost at the shore and the other reaching outward, to deflect the schools of fish and funnel them into the pocket, as the enclosure is called. Seen from a hill above the water, a weir with its long, graceful leaders and circular pocket is a thing of geometric beauty. Usually it is built well up some inlet, though it sometimes is constructed off a ledge or small island in more open water.

A landowner has all control of weir fishing off his shore, so that the ownership of tiny islets and sterile ledges is guarded zealously by those who know enough to see in them promising sites. If anyone wishes to build a weir in water fronting on someone else's property he must make arrangements with the owner. Anyone building a weir, even off his own shore, must procure a license from the local selectmen, who will issue it only if they are convinced that the proposed weir will not obstruct navigation or in any way interfere with the rights of others. When a weir is no longer fished, the owner must remove the stakes and piling, which are rightly considered a great menace to boats.

Weir operating is perhaps the greatest gamble in all the fishing on the Maine coast. The only mitigating factors are that the ante is not great and the winnings can be enormous. Unless he builds near some off-shore islet or ledge, the weir operator doesn't even need a power boat. A couple of dories are sufficient to row out to the trap and seine it when a catch is made. But he must own or hire a piledriver, which for this purpose is usually a crude affair mounted on a small leaky scow and operated by a hand windlass. The pilings, stakes, and brush he can get from his own or a neighbor's woods for nothing. But he must be prepared for a lot of work with ax and saw. As one man put it, "When you build a weir you first cut twice as many poles and brush as you think you'll need. Then you go back and cut as much again. Finally you spend three more days in the woods and you may have half enough to build your weir." Here and there along the uninhabited shore the hiker comes upon an acre or so of recently cleared land bearing witness of these Herculean undertakings. The only money expense is for the "twine"—the net used in the pocket for seining out the catch once it is trapped there. Usually a sardine cannery will supply the seine if the fisherman agrees to sell his catches to that company. A small percentage is deducted from the payment for each catch to pay for the net.

Every weir fishes differently. Some are good at high tide, others at low. Some do best in spring, others in autumn. Some produce in a west wind, others in an east. Each one fishes according to its own pattern and that pattern is not the same from year to year. The majority never do well enough to justify their construction. Some have one or two profitable seasons, and a few do consistently well for many years. The owner needs patience but, even more, he needs some other means of livelihood while he's waiting for his weir to produce.

All along the coast are stories of weirs that in two or three nights have made fortunes for their owners. In the most impressive of these stories, the weir has been a dud for years. Then one morning the owner goes out to find the pocket bulging with squirming silver. The trouble with a catch like this is that the run may have been phenomenal elsewhere and a dozen other fishermen the same night may have made similar catches so that the market is flooded and prices depressed. But when such hauls occur in isolated cases, the fisherman makes a killing indeed. Ruth Moore in *The Weir* captures realistically the drama of this situation. A fisherman who for decades has been wasting time and money on a weir that never pays off in more than an infrequent bushel or so of bait for the local lobstermen goes out one morning to find his trap loaded. Two sardine boats from rival companies appear and bid up the price. Before they leave, the fisherman by his own modest standards is a rich man.

A weirman lives by the hope for the "big day," or rather night, when for reasons beyond the understanding of man the herring pour by the millions into the pound. Swans Island has had weirs that in a single year have caught as much as $40,000 worth of fish—$20,000 for the owner and $10,000 to each of two helpers—and in the following year have made nothing. At nearby islands weirs are reputed to have made $60,000 in a season.

Many weirs are in remote coves away from roads and settlements. The owners of these build little cabins in which to cook and sleep and above all to wait—especially at night, for that is when the herring are most likely to school inshore. Almost every lonely inlet or island has one of these shacks—either standing or in ruins—as the monument of a venture entered upon with vast hope but all too often ending in failure.

The great advantage of the weir is its cheapness to build and operate. Its great disadvantage is that it is stationary. It must wait for the fish to come along according to their own undependable nature. Thus the weir has slowly been going out of favor and is being replaced by seines, which can be taken wherever the fish are. But seining requires a very sizable initial investment. The sardine companies will help finance the great nets, which are much more expensive than those used in a weir. Spread out in a field to dry, one of them seems to cover acres. A large boat is also needed for carrying them from place to place and for hauling the dories used in setting them. At least three men are needed to operate a seine.

On Swans Island at present are a number of seiners. One group of four men uses two large boats, an outboard motorboat, and a big dory or seine boat. The largest boat carries equipment and provides living quarters for the men when they are away from home. The smaller of the two inboard boats scouts the water for miles about for herring schools. Much of the scouting is done on moonless nights when the fish set the surface of the water aflame with sea-fire. Formerly the scout had to depend on the sharpness of his eyes to see the fish, though in the daytime a clue to their whereabouts is given by the gulls circling over a school near the surface and by the presence of seals and porpoises. In recent years, however, radios and fathometers have vastly simplified the job. With a fathometer the scout, day or night, can locate fish that may be too far beneath the surface to be detected by the eye of man or gull. The schools of fish are so compact that they effect the instrument just as a bottom does and thus give away their position to the fisherman who by glancing at his chart knows the depth of the actual bottom.

When the scout boat locates the fish it radios back to the others, who set out at once in the big boat with nets and with the dories atow. When they arrive where the school is, the outboard hauls the seine dory as the men attempt to encircle the fish with the twine. If they are successful and the fish don't escape underneath or through the jaws that are enclosing them, the men purse up the bottom and the fish are captured. The process requires little time but the utmost skill. If the catch is small, the men load it in their own boat and very likely sell it to the lobstermen, ever clamorous for bait. If large, they may radio a nearby sardine factory, which will dispatch at once one of its boats to empty the seine with the aid of a huge dipnet slung from a crane. These sardine boats are always on call in case a seiner or weir makes a big catch. The canneries have an insatiable hunger for this tiny fish.

The great enemy of the purse seiner is fog, which makes it impossible to spot the schools, though the fathometer is a help against this handicap. Sometimes among these eastern islands there are as many as a dozen days of fog in a single summer month, during which the seiner can do little more than bite his nails and develop ulcers. Another type of twining is to some extent free of this drawback. It consists of shutting the herring off with a net in a cove where they habitually go. The net is essentially nothing more than a portable weir taken to the particular inlet or bit of shore where herring happen to be. These spots which over the years these fish have been known to favor can be scouted in any sort of weather.

Walking at night in the fog along an island road where it skirts the shore, one sometimes finds himself in the midst of these delicate operations. First he hears the herring themselves. Above the gurgle of the tide among the cobbles he hears a sound like the sizzle of summer rain on an inland pond. The whole shore bubbles and seethes, though the fog is too thick to see the water. Soon quiet voices draw apart in the fog and then approach one another again. Suddenly an outboard motor snarls, then buzzes out of hearing across the shrouded water. The fish boil more violently among the rocks and seaweed at the edge of the tide.

All during the night, light sleepers in the shore houses are aware of unwonted doings. The oars continue to rattle, the outboard snarls spasmodically, and voices drift in cheerily in contrast with the dolor of the Duck Island horn, which is customarily the only voice on the sea on thick nights. Sometime before dawn the wind changes, the fog gallops out into the ocean, and the wind drops off. The stars come out and the northeast begins to pale, then reddens and bursts into flame. The water absorbs the glow and burns faintly beneath a turquoise sky where only a few stars linger in nacreous pallor. From their windows the villagers see the results of the night's work. Across the mouth of a shallow cove a little up the shore is a line of cork buoys, like a necklace curving slightly outward. Beneath the necklace, they know, hangs the "stop seine" drawn by its leaden weights almost to the bottom. The water of the bay is ruffled by the merest catspaw of a breeze, but inside the string of corks is a pond of glass above which gulls, rosy in the first rays of the sun, wheel and gabble excitedly. Once in a while a bird plunges into the still water behind the net and flaps up again with wriggling silver in its beak. The villagers become tense. Flocks of gulls over a weir or seine mean a big catch.

In the deeper water off the cove is anchored the seine tender, the base of operations. Near the middle of the string of buoys the big dory is being slowly towed by the outboard motorboat in an arc that extends outward and then bends back to meet the buoys again. In the dory two men are vigorously paying out the twine—that is, another net—which as it sinks forms its own necklace looping outward from the other. Into this net the herring will be driven through an aperture opened in the larger one.

Meanwhile around the point at the northern extremity of the island a white sardine boat from Mt. Desert Island has appeared. The tide is low and the boat, with mast and crane jib gracefully aslant, rounds all the black buoys so as not to go aground. Cautiously she feels her way in until she casts anchor a hundred yards or so from the net. The skipper climbs into a punt and rows over to the seiners. A consultation follows; then he rows back to the boat. The seiners now go in behind the main net and drift about banging their oars on the gunwales and splashing them in the water. The herring rush for deep water but succeed only in pouring into the smaller net, the pocket, where they are trapped. Soon with the rising tide the sardine boat will draw in closer and with her crane and great dipnet will start taking the catch aboard.

But seine fishing, almost as much as weir fishing, is a gambler's game with all the excitement, disappointment, and eventual near certainty of loss that gambling involves. On Swans Island most of the weir fishermen have other more reliable ways of making a living, usually lobstering. The seiners have so heavy an investment that they have to work their equipment as near to full time as possible. Yet they, too, often run a string of traps when time permits. One thing is certain. There is always a market for herring when they are caught. And there is always the chance for the big killing—a better chance than with a weir. The canneries will always welcome the large catches and the small ones will be snatched up avidly by the lobstermen, who will descend like a flock of gulls at the first rumor of a catch.

The bait problem is perennial with the lobstermen. If they can't get it from the local weirs and twiners, they must go elsewhere. Often they cross the bay to purchase the waste from the canneries and fish-packing plants. Sometimes, as was their habit in former days, they are forced to go handlining for pollock or spearing flounders for bait. But they would much prefer to buy fresh herring from local fishermen, and they are willing to pay well for the chance. Time spent in searching for bait is time lost for the lobsterman. Thus the herring fishing on Swans Island is in some measure dependent on the lobstering.

5

The Lobster

TODAY THE COAST OF MAINE LIVES BY THE LOBSTER, WITHOUT which villages would fall to ruin and the shores would become the undisputed domain of seals, gulls, and summer cottagers. This crustacean's range is from Delaware to Labrador. But at present for all practical purposes it is from New Hampshire to Labrador. South of Kittery he has ceased to count as a commercial fish. F. H. Herrick, the author of two definitive book-length treatises on the lobster describes as follows the evolution of lobster fishing in any given area: "1. The period of plenty, when lobsters are large and abundant, the price low, and the fishermen few; 2. The period of rapid extension, when price advances slightly and the fish yield declines a bit; 3. A real decline, with fluctuating yield, but more fishermen nevertheless, and with still further advancing prices; 4. A period of general decrease—in everything except price."

In colonial times lobsters were plentiful in New York Harbor and around Hell Gate. But as the city grew, what few lobsters the fishermen didn't capture were killed by water pollution. For a while Cape Cod was the chief source of supply for both New York and Boston, but by 1812 these grounds were beginning to decline and by 1880 had all but vanished. Gradually the intensive fishing worked eastward along the coast, depleting the waters as it progressed. Even as early as 1870 American capital was opening lobster-meat canneries in the Maritime Provinces because of signs of serious decrease in the Maine supply. By 1935 imports of fresh lobsters from Canada by refrigerator car and smack equaled the American catch. At present Canadian imports are even larger, and the northernmost limits of the lobster's habitat in Labrador and Newfoundland are being tapped. The Maine yield, thanks to vigorous conservation measures, has leveled off, temporarily at least, at about an eight million dollar industry employing approximately five thousand men.

Till the middle of the last century, no lobster traps had ever been set in the waters around Swans Island. Frequently while trawling, fishermen would pull in a lobster on one of their hooks. But, as a general thing, if a housewife decided to have a lobster dinner she would send one of the children down to the shore at low tide. Poking around among the rocks and peeking under the mattings of sea weed along the beach, the child would have as many three- or four-pound lobsters as he had been told to get. This was the aboriginal way of catching them, the one used by the Indians and the early settlers from Delaware northward.

Then in the autumn of 1850 a smack with four men and a load of traps aboard arrived at the island from Gloucester. Hiring several islanders to help them, the men remained till the tanks of the smack were full and then returned to Gloucester. At Swans Island had begun the occupation which a hundred years later is supporting almost every man, woman, and child in the community in greater prosperity than it has ever known.

Expeditions like this had been made ten years earlier to the western stretches of the coast. The purpose was to show the inhabitants how to catch the lobster so that in the future they would supply the dealers. Until the visit from Gloucester the Swans Islanders had had no idea of the abundance of lobsters in their waters. The Gloucester fishermen never returned, but the next year a number of islanders built themselves some traps and tried their luck.

By the summer of 1855 ten men were engaged in lobstering, and smacks from Boston were making monthly calls to pick up the catches. These men, who were usually too old to go on long fishing voyages, set thirty or forty traps apiece. Their daily catch would be from two hundred to two hundred and fifty lobsters, each one from three to three and a half pounds in weight. The 1850's at Swans Island were Herrick's Period One. Lobsters were large and abundant and fishermen few, and prices were so infinitesimal—a small fraction of a cent a pound—that even the ten old men gave up for a number of years after 1855.

By 1860 the second period had set in, that of rapid expansion with slightly higher prices and slightly reduced supply. The establishment of canneries along the coast provided the chief stimulus. For a time a cannery on Swans Island induced many fishermen to put out a string of traps. However, the supply still generally exceeded the demand and prices remained discouragingly low. But by 1870 factories in half a dozen neighboring towns, not to mention the smacks that in increasing numbers were clamoring for business, were offering good prices. Ten years later the third period, that of "real decline, fluctuating yield," more intensified fishing and much higher prices had begun. In 1879 seventy-four men—about as many as at present—were trapping lobsters on Swans Island. About fifty of these trapped only in the spring and early summer, and about mid-July went to sea for the more lucrative mackerel seining. The remaining twenty-four, mostly elderly, lobstered on during summer and fall, except for a lay-off during the shedding season in August and early September. During this period when the crustaceans are discarding their old shells and growing new ones their meat is inedible.

By this time the average lobsterman on Swans Island was

working about twice as hard as in 1855 and getting about half the catch. But he was making more money. He was setting about one hundred traps; owned two boats, a large one for hauling traps and a small one for catching bait; and was capturing each day somewhat under 200 lobsters of from one to two and a half pounds each. The larger ones he sold to smacks for the fresh market and the smaller ones he sold to canneries, generally to one at Deer Isle, for the Swans Island cannery had soon gone out of business. If he had a good year he would net three hundred to four hundred dollars for about seven months of fishing. The average for all of New England was only about one hundred and forty dollars for each man, though many did not fish a full season. The average price per pound was two and a half cents. Yet in 1879 the lobsters were plentiful as compared with the present. Some of the best trapping was done so close to shore that the pots would be out of the water at low tide. The lobsters seldom "work in" that far today, presumably because they are less numerous and can find sufficient food below low water mark.

From 1890 onwards the island specialized more and more in lobstering, mainly because there was little else to fish for with profit. Soon as many as one hundred and forty men were setting traps, with a total investment in gear and boats of more than thirty thousand dollars. The yearly catch was worth about seventy-five thousand dollars, or over five hundred dollars gross and over four hundred dollars net for every man fishing. The price per pound had climbed to nearly nine cents. This is the long-range pattern—more fishermen, fewer lobsters caught by each fisherman, but a sharp enough rise in price to make the occupation continuously more profitable. The tendency held till around 1910, the total catch remaining about the same for the state as a whole. Then just before the First World War there was almost a fifty percent drop within a few years in the state-wide catches. Perhaps this was the beginning of Period Four—general decrease in everything but prices.

Between the wars periods of hard times occurred. In the mid-twenties prices declined when Canada started shipping in lobsters, duty free, from as far away as Cape Breton and Anticosta Island, and even Newfoundland entered the competition. The Great Depression was severe on the islanders as on the rest of the nation. Lobsters dropped from almost fifty cents a pound in 1929 to about twenty in 1930, though operating expenses—gasoline, bait, equipment—dropped but little. By 1935 the price had sunk to fifteen cents a pound and had climbed to only sixteen cents by 1941. During these years the island lobsterman was lucky if he grossed an annual five hundred dollars. Even that was more than the state average, because the islands were in the midst of the best fishing grounds. Some men gave up lobstering to seek a living elsewhere, but most hung on, knowing they would always have at least enough to eat. On the island they owned their own houses, could cut their fuel supply in the forests, and could raise a few vegetables and get all the sea food they wished. To those who stuck it out the buyers extended credit, in some cases as much as three thousand dollars, for gasoline and gear.

2

It has been suggested that after Period Four a fifth period occurs—one in which the people through their government try every thinkable legal and scientific measure to preserve an important means of livelihood. Closed seasons, size limits, licensing of fishermen had long been resorted to. Now they were enforced with greater rigor. A law was passed to protect seed lobsters—the females with myriads of eggs clustered beneath their tails. Hatcheries were established and lobster fry scattered along the coast. As early as 1898 Swans Island received two hundred thousand fry. Gulls, which under the protection of federal laws were multiplying rapidly, were found to eat the helpless larvae of the lobsters. So state men visited the nesting grounds to sterilize a portion of the gull eggs by pricking them or by spraying them with poison. To protect the native fishermen licenses for trapping lobsters were issued only to persons of more than five years' residence in the state, an effective way of keeping summer cottagers from cluttering the water with traps. Federal authorities, responding to the indignant protests of the fishermen, ruled that lighthouse keepers, who had been finding time to set sizable strings, were limited to thirty traps each.

Better times came with the Second World War and lingered with the inflation. The lobster business as it is today is definitely a war baby conceived the moment the price control board designated lobster as a luxury item not subject to ceilings. Very soon the price leaped to seventy or even eighty cents a pound in winter, and for ten years it has not gone below thirty cents even for shedders. At the end of the war veterans returned to find the half-dead villages they had left a few years earlier aglow with prosperity. Throwing themselves into the business with abandon, they purchased licenses and tossed together a string of a hundred or so traps. They caulked and painted old boats drawn up years before on the banks to die, equipped them with new motors, and launched them for a new lease on life. In many cases they built or purchased new, bigger, and more modern boats, often on G.I. loans or other forms of credit. For the state as a whole the number of licenses issued increased almost fifty percent over pre-war years. If the families of fishermen and all those employed in buying and transporting lobsters are counted, eighteen thousand people in the State of Maine have been living off this fishery in recent years. At least 90 percent of the population of Swans Island would be included in the number.

On the island the average number of traps for each man rose to nearly two hundred. A few had out as many as four hundred. Old-timers who considered one hundred traps the most efficient number were filled with wonder and chagrin. One lobster-buyer summed up the situation as follows: "To fish over one hundred or at most one hundred and fifty traps is stupid. It's wasteful and it's a sign of greenness. You can't take care of two or three hundred traps the way you ought to. Also, it shows a disregard for the lobster supply. How long will there be any lobsters if they go on fishing them that way? It's sheer slavery, too, with that many traps—on the water at four A.M. and not back till six or seven in the evening. Where's the chance to enjoy your family? You're so damned tired when you get back, all you can do is eat and crawl into bed so's you can get up at four the next morning."

The dealer is right on all counts. It is hoggishness and shortsightedness to fish so many traps. Most men realize this and are hoping for legislation limiting the number. Around the whole island and among the neighboring islets the belt of buoys and toggles is so thick that a boat can hardly squirm through them without fouling a warp. "Formerly," a lobsterman explained, "if catches got poor in one place a man could move his traps to another. Now there are so many traps that there is no place to move them to. You're lucky if you can find a few feet of clear surface anywhere." If a man finds a spot where fishing is good, he will, of course, not advertise it, but he won't keep it long for himself. One buoy attracts others.

To make the lobster's life even more hazardous, many of the fishermen have fathometers on their boats, electronic devices which at the press of a button tell the depth of the water. Lobsters feed on shoals or around the edge of them.

Formerly these bottoms had to be located by soundings and landmarks. The push-button is surer and quicker. The time saved can be used in setting more traps.

Where competition is so fierce trouble is sure to flare up. In 1950 the Commissioner of Sea and Shore Fisheries was "sorry to report" twenty-eight cases of gear being molested in the state as a whole. In each case the malefactor's license was suspended. Very laconic! For each of these cases prosecuted by the Commission, there were untold others settled in different ways, sometimes involving a long and bitter quarrel. A lobsterman whose catches are poor or whose traps disappear becomes suspicious that someone is hauling his gear or cutting off his buoys. Around the wharves he starts muttering darkly that he knows what's going on and who's doing it. Often he is simply the victim of bad luck. The lobsters have boycotted his traps and the propellers of other boats have inadvertently fouled his buoys. But sometimes his suspicions are well founded. In reprisal he may haul his suspected persecutor's traps, or he and a friend or so as witnesses will watch with spyglasses from the woods at dawn to catch the thief in the act. Frequently there is much braggadocio about the use of guns; and occasionally cankering feuds, "lobster wars," develop, though the threats of violence are seldom carried out. Sometimes the issue is settled by the local sheriff or constable. But more often the fish wardens are brought in and once in a while a case gets to the courts, as happened with the lobster war around Matinicus in the summer of 1953. No heavily fished area is without this sort of thing. Swans Island has had its share.

A further cause of feuding is the intrusion of fishermen of one community into waters habitually fished by those of another. Legally any licensed fisherman can set his traps anywhere in the state. But there is a tacit agreement among neighboring communities to stick to their own waters. When the line is crossed, wholesale cutting of buoys on either side may occur, but generally the outsiders get the worst of it and withdraw, almost always without any appeal to the law.

The effect of all the activity on the lobster grounds has been much the same as if in raking your lawn you suddenly used a rake with teeth a half inch apart instead of one with teeth three inches apart. You would start gathering great quantities of trash that all previous combings had left. Thus at present about five thousand men in the State of Maine with half a million traps and forty-two hundred powerboats rake up from the bottom twenty million pounds of lobsters worth eight million dollars, a catch that in size equals those previous to 1910 and is four times as great as that of 1935 in quantity and nine times as valuable. But the catches of 1910 and 1935 were made by fewer men using much less expensive equipment.

Furthermore, since the Second World War fishing has been much more continuous throughout the year than it formerly was. It is the habit of the lobster to spend the winter in deep water well off shore. In the spring he slowly creeps in until by midsummer he is in depths of only a few fathoms. In August and early September he sheds and is not fit to eat. After the growth of his new shell, and with the approach of cool weather, he works back again to his winter quarters. Years ago lobstermen fished about four months in the spring and summer, stopped for a month or six weeks in the shedding period, and then fished an autumn season of two or three months.

But since the war men have been pairing up in the larger and more seaworthy craft and keeping traps out all winter in the deeper water. In winters of persistent winds a boat may not get out more than twice a month, and always winter hauling is risky business, even with two men aboard. Yet it brings in just so much more money. After a slump in December when the season is open in Canada and there is a flood of lobsters from the Maritime Provinces, the price soars to sixty cents or more. So long as there are any lobsters to catch, the temptation is great, for such returns, to go after them even at personal danger and discomfort. The end result is to swell the huge annual catches and put even further strain on the already endangered supply.

The phenomenal thing is that these catches have been going on since 1947. It is difficult to decide which deserves greater credit: the stubbornness with which the lobster keeps his hold on existence; the energy and skill of the fishermen; or the State Commission that has labored so indefatigably to conserve this resource. One thing, all fishermen agree, has helped immeasurably: the laws are more strictly enforced. A man can be heavily fined and lose his license for keeping short or seed lobsters. Fifteen years ago one could often find bags of shorts tied to the spilings of wharves used by lobstermen; later these illegal catches would be sold or used by the fishermen themselves. But a more numerous force of wardens armed with heavier penalties has almost wiped this practice out. The fishermen, always haunted by the spectre of a lobster shortage, have also come to realize that their living depends on the fishing laws. Some are advocating more far-reaching legislation. On Monhegan the fishermen, in an effort to save the lobster supply, have successfully petitioned the state to enforce a closed season around their island. Eventually other communities may have to do likewise. But too often islanders try cooperative action of this sort only as a last resort, when it may be too late.

3

What has this lobster bonanza meant to Swans Island? A rough estimate for the total annual catch would be about five hundred thousand pounds or about one-fortieth of the entire

Maine catch. The total receipts of the fishermen for five hundred thousand pounds of lobsters at an average of thirty-five cents a pound would be one hundred and seventy-five thousand dollars. As about sixty-five men are fishing, the gross take for each fisherman is around twenty-five hundred dollars, which despite inflation is far above anything Swans Island has ever known before.

All of this is not profit, of course. The gasoline and oil for a year's lobstering run to three or four hundred dollars. There are repairs and upkeep on the boat and engine, themselves an investment of a couple of thousand dollars which depreciates with every year of use. Each trap of the fisherman's several hundred costs five to ten dollars, depending on the depth at which it is to be fished, that is, on how much rope, or warp—a very expensive item—has to be used to moor it. Under the best of conditions traps wear out and warps grow rotten and have to be replaced. Barring accidents a fisherman has to count on the renewal of his complete string of traps every two years. The claim is that a trap will pay for itself in a week's good fishing, but more frequently it takes longer. In any event this constant replacement of gear is a major expense. The lobsterman is most lucky who can count as much as 75 percent of his gross as profit. And this doesn't allow for the big setbacks, the acts of nature that plague the fisherman as much as the farmer. For example on Labor Day of 1953, a small hurricane struck the Mt. Desert region. Beginning at ten in the morning Swans Island was swept by gales up to seventy miles an hour and its eastern shores were pounded by huge seas. At that time of year the traps are almost "in the woods." The undertow of a heavy sea kicks them around on the rocky bottom like scraps of paper. Hundreds of traps were splintered or washed ashore a hopeless shambles of broken laths ensnarled in rope. That same autumn there were two other equally destructive "breezes." Storms are common on the coast and no fisherman can expect to wholly escape their damage. One gale can cost him a thousand dollars.

All have not shared equally in the big catches of recent years. The incomes from lobstering on Swans Island range from a hundred to ten thousand dollars a year. The low incomes are made by boys and old men puttering around the coves in rowboats. A middle group of serious, full-time fishermen with moderate-sized power boats take in from twenty-five hundred dollars to five thousand dollars a year. The big incomes are made by owners of large boats—up to forty feet in length—who fish with large numbers of traps and venture farther into the deep water than their competitors. Their powers of work and endurance, combined with a shrewd boldness, not recklessness, and an intelligence far above the average hurtle them on inevitably to mighty successes. In the old days men like these commanded the high-line cod and mackerel schooners like the *Alice*. They are open and talkative, above the provincialisms of their fellows. They speak with quiet scorn of those fishermen who stay close to shore. Summer and winter they range the open ocean where the "big ones" are, for even in summer many lobsters remain in deep water. Two such men from the mainland, who confine their fishing to the waters off Mt. Desert Rock, twenty-two miles from the nearest harbor, are said to net as high as fifteen thousand dollars a year. Islanders whose incomes are in, or approach, five figures keep a Buick or Cadillac ashore for pleasure use, and a Ford, Chevy, or Jeep on the island for batting around. To keep their income taxes within reason, some of these oil-skinned tycoons fish for only part of the year. When they are not working they may vacation in Florida or tour across the continent in a luxurious vehicle paid for in cash extricated from grimy rubber boots before an astonished salesman in a Rockland or Bar Harbor showroom.

Beneath these pinnacles are various gradations of success.

In its simplest form lobstering is done from a rowboat, probably a peapod—a light, low-sided, double-ended, round-bottomed boat, which was invented among the islands about 1870. The very young and the very old are the chief users of peapods. Boys of twelve or upwards employ them to tend a string of traps which they set partly for the fun of it and partly to earn pin money, or to help pay their expenses if they plan to go to high school ashore. Old men without great needs and without ambition and an occasional young man who is simply shiftless will use the peapod.

A man or boy fishing from such a boat can handle not more than fifty traps. He does most of his hauling within a mile or so of his mooring. As you enter a Maine coast harbor you usually see several red-faced lads and grizzle-faced oldtimers in battered oilskins that hang on them like rockweed on a ledge at low tide, "pushing" their peapods among the buoys off the harbor heads. The advantage of this type of rowing from a standing position, with short, highly feathered sculling movements of the oars is that the lobsterman faces in the direction in which the boat is moving and can thus guide himself easily to his buoys. The peapod fisherman doesn't get rich, certainly, but does well enough at times. He has no operating expenses in his boat, which represents a negligible investment. His catches are small but he clears more on them than the larger fishermen. His main handicap is the restricted area he is forced to fish.

Most lobstering is done from inboard-powered craft about thirty feet long. These boats have a "house" with glass across the front and entirely open at the back, and with a roof projecting sternward well beyond its narrow sides. In the house is the wheel, the instrument panel, and a winch and pulley on a davit for hauling traps up. In front of the house is a sprayhood or a decked-in space forming a cuddy. Here are the engine and equipment such as life jackets, fuel, and oil. If the boat is to be used for winter fishing there is also a tiny wood or coal stove to keep the lobsters from freezing, and to provide a refuge of warmth for the men and means of brewing a pot of coffee. Along the sides of the boat behind the house are narrow washboards with a high comber and numerous scuppers. In the stern is an ample decked-in space for other equipment. The central part of the boat, the "standing room," is undecked but is provided with a platform or flooring about a foot above the keelson. The fisherman works and stows traps, bait barrels, and his catch here.

Most of the boats were formerly powered by second-hand automobile motors. No engine was too old for use in a lobster boat. In the summer of 1941 the author had on the island a Model A Ford that for twelve years had piled up vast mileages rattling back and forth across the continent. Most of the time on the island the vehicle refused to run; the fog, which crept into its ignition, was deadly to it. Yet at the end of the summer he found an eager buyer for it for the price of twenty-five dollars. Keeping these derelicts going may have stimulated the mechanical ingenuity of the islanders, but it did not make for efficient lobstering. So with the flush times after the war most fishermen installed marine engines, usually Chryslers or Grays straight from the factory, and time wasted in "tinkering" has notably decreased.

Every fisherman considers himself a final authority on boat design, and has his own deep-seated notions as to what makes an able craft. From babyhood upwards he has heard boats discussed, as have his ancestors for six or seven generations. Much lore, ranging from sound knowledge to superstition, has been handed down—for example, one shouldn't change a boat's name or have any blue paint on her. What every fisherman hopes to do, and frequently succeeds in doing at least once in his lifetime, is work with a master boatwright in building his own boat. This kind of partnership has turned out the perfect craft for many a fisherman. The two will work a whole winter pondering over plans, building models, and finally fitting together the planks and timbers. The result is something as dear to the owner as life itself, a sweetheart to be petted and coddled for many more years than any human sweetheart could expect. The fisherman's whole family shares his pride in his boat. Deep in the make-up of all sea-faring people is a reverence for a boat as almost a living organism—a feeling that goes beyond mere considerations of making a living.

4

The lobsterman's day is a full one. If you are a late sleeper —that is, if you lie abed till five A.M.—you will be aroused by the splutter of one boat after another as she leaves her mooring and coasts out the harbor to the first trap buoy. Perhaps, if you were half awake before that, your dreams were troubled by the rumble of punts being dragged down the beaches and of oars squeaking in rowlocks and banging on wooden bottoms.

In recent times you would also hear spurts of music, the stentorian bellowings of news commentators, and electrically magnified conversations wafted across the water and through your window from the radios, ordinary and ship-to-shore, that are fast becoming standard equipment. One summer visitor who had traveled almost a thousand miles to enjoy the wilderness quiet of the Maine coast complained of being awakened in a sweat of terror at five one morning by the full blast of a jazz band issuing from the loudspeaker of a boat hauling traps just below his window. As the day wears on and more boats are on the water, the conversation on the two-way sets becomes long-winded and boisterous. The air-waves are cluttered with horseplay as fishermen pretend to be calling from

George's Banks or Europe. The Coast Guard has occasionally given warnings to the most inveterate ether hogs, and old-timers who don't have sets mutter sourly about the desirability of passing laws and not seeing the use in the durn things—just a lot of damned nonsense sent over them anyway.

By the nature of their calling fishermen are gregarious people. Aboard ship they are constantly in the company of their fellows, doing work which requires a high degree of team work. On shore they spend long hours on the wharves and in the stores talking shop and spinning yarns. Lobstering goes counter to this deeply ingrained tradition in that it is usually done by a man alone in a boat. The two-way radios do much to restore the sociability of the "bankers" and seiners.

There is monotony, as well as loneliness, in the lobsterman's work. From dawn till far into the afternoon, if he has a large string out, he repeats over and over again one set of motions. Threading his boat among the clusters of buoys which, wherever there is promising lobster ground, bloom in every color of the rainbow like patches of miraculous water lilies, he searches out whatever identifying color or combination of colors in rings or stripes he has painted on his buoys. When he finds one he pulls alongside it, bow to windward, and with a flip of the gear lever transfers the engine power from the propeller to the winch. With a gaff he hooks up the warp, places it on the pulley on the davit overhanging the water, and gives it a twist or two around the winch. When the trap breaks the surface, he reaches over the side for it, at the same time slipping the warp from the winch, and balances it on the washboard. The trap is a crate of laths nailed on flat sills of hackmatack or spruce to form the bottom and spruce boughs bent to form the rounded sides and top. One end of this half-cylinder, which is three feet long, two feet wide, and one foot high is covered with netting funneling sharply inward to a small opening through which the lobsters, attracted by the bag of bait hanging inside, may enter but not exit. An interior funnel of netting makes a second compartment in which the lobster is eventually lured, thus making his escape impossible. Along one side of the trap is a lath door hinged with leather straps and fastened with a wooden button. In the floor of the trap are fastened stones or pieces of iron to keep it from floating. During the winter in his own shop the fisherman has carved his buoys from chunks of wood and built his traps from island wood sawed at his own or a neighbor's mill. The heads, as the funnel-shaped nettings are called, he may have bought ready made, but if he is particular he and his wife have probably knitted them out of twine.

When he has the trap resting on the washboard the fisherman opens the door, flicks out the many crabs—often dismembering them and taking the claws home—and short lobsters that have been feasting off his bait and off each other, measures the doubtful lobsters, and tosses into a tub those that are right, sometimes plugging the claws with wooden or plastic pegs. He is lucky if he finds one keepable lobster in each trap. Next he detaches the bait pouch, which hangs in the "parlor" inside the second netting, throws its stinking contents to the cloud of gulls that escort him from buoy to buoy, and inserts a pouchful of new bait from the barrel at his side. The bait is half-decayed fish that has been gathering strength in pickle—that is, in salt—for a number of days in the mephitic depths of a fish shanty. (The lobster is believed to depend entirely on his olfactory powers to locate food.) Once the new bait is installed, the fisherman engages the propeller and tosses the buoy overboard, which is quickly followed by the toggle—an old whiskey or beer bottle or a block of cork attached to the warp shortly below the buoy to keep the warp off the bottom at low tide. As the line pays out the fisherman scans the water for an open place for his trap, no easy job in these days of crowded fishing. When he finds space he slides the trap into the water and with a burst of speed bears down on the next buoy. Repeat this series of moves from one hundred to four hundred times, and you have the lobster catcher's working day. When the last trap is hauled he returns to his mooring and stores his haul in cars—slatted crates of heavy wood afloat in the water—or he goes directly to his dealer who weighs out the catch, always on the lookout for shorts and shedders, and hands him the money for the day's take.

Yet the work has its attractions. To enliven it there is the element of luck, of uncertainty, that is the fascination of any fishing. You never know what the next trap will bring up. Many a time a man hauls in his supper in the form of a fat cod or cusk. Also, the weather and the sea are never the same from day to day, or from hour to hour. Only the thickest fogs and heaviest rain keep the fisherman home in summer. Often the day offers a challenge to his seamanship, and often the protean beauty of the water, coast, and sky gladdens his labor. There is scope for intelligence, too, and for that sixth sense that the successful fisherman must have. Lobsters do not always remain in the same place. Aside from their seasonal crawls in from deep water in the spring and back again in the fall, they move laterally from spot to spot alongshore. To detect these movements the lobsterman must always have a few experimental traps out in places other than those in which he is doing the bulk of his fishing. Above all he must have experience. He must know every square foot of bottom —whether rock or mud, the tidal currents on it—before he can set his string effectively. In lobstering more than in most other occupations, initiative, energy, and intelligence are the requisites of success.

6

Work Ashore

ONCE THE MAINE COAST FISHERMAN WAS AS MUCH AT HOME IN the woodlot, the fields, or the cowbarn as on the water. As a farmer he might have had a flock of chickens, a few sheep, a cow or two, a yoke of oxen, or a team of horses. On the wharves of Bangor you were just as likely to hear the schooner captains discussing their potato crops as the state of the tides.

On some of the islands as well as on the mainland, farming, fishing, and seafaring were often combined into what must have been a stimulatingly diversified existence. A stirring description of such living is given in Charles Eliot's account of John Gilley, who fished and farmed on Baker's and later on Cranberry Island, both only a few miles from Swans. Indeed, the old way of life lingered somewhat longer on the islands than on the coastal mainland. Eliot praises the self-reliance, the versatility, and the alertness this life of many occupations promoted. Very likely the so-called Yankee virtues were more highly developed among the people of the Maine coast than elsewhere in New England, or at least they persisted to a later date.

Some islands had an extensive agriculture in the past. Matinicus, which is as flat as an Illinois prairie, has many typical Maine farmhouses with barns as big as those on the mainland eighteen miles away. The level fields were once highly productive and the pastures on the headlands and neighboring isles provided lush grass for large herds of livestock. Even today on nearby Ten-Pound Island a small flock of sheep graze summer and winter with no fresh water and no shelter, keeping the surface of the island as close-cropped as a golf course. It is a strange sight twenty miles out in the Atlantic to see the flock on this treeless, wind-swept island, their heads all pointing in one direction like the bows of a fleet of workboats at their moorings. Some of the other islands have had little agriculture. Isle au Haut and Outer Long Island had only a few farms of any size, partly because of rocky, hilly terrain and partly because the men were too busy with the sea to take on more than a garden patch and fields enough to support a family cow.

Swans Island at one time had a number of farms, but at present has only two or three that come anywhere near supporting their owners. The fields never were so broad nor the barns so spacious as on Matinicus. The land is rockier and hillier. Of a dozen or so barns still standing, some are dragging out their last days with the aid of spruce pole buttresses. The interior of the island is the site of barns and farmhouses of which only a few shattered timbers remain in briar-grown cellar holes. The huge granite foundation slabs, hewn from nearby boulders, are covered with lichen and heaved out of place by generations of frost. Nearby are decaying orchards, the trees long since gone wild and now half dead, and clumps of lilacs, those ineffaceable and ubiquitous symbols of the American homestead. The old fields are sometimes kept partly clear by the pasturing of a steer or two on them, but the stone walls that devoured the vitality of so many men and oxen in building have long since collapsed into shapeless heaps.

If you search a bit you will find a family burying plot, its headstones prostrate in the deep grass, as forgotten as the farm where those who are buried there once lived. All around, the alders and the spruces, those tireless skirmishers of the wilderness, are creeping in. Among the dying branches of the apple trees the wind rustles complacently. Down a vista among the hills sparkles a patch of sea, and now and then the breeze carries up the dying note of a bell buoy. The scene is steeped in the peace of resignation and defeat.

2

With the island's growing specialization in lobsters, less and less farming or gardening of any sort was done. The use of bigger boats and the development of year-round lobstering made the smaller coves less usable. The roads that once threaded the interior of the island grew up in grass and brush. The part of the island west of the Carrying Place, where there had been much farming but poor anchorages, reverted quickly to woodland. Today scarcely two hundred acres are pastured or cultivated, about a fifth of the former tillage. Livestock has dwindled to only three horses, two oxen, a dozen or so cows, and ten sheep. The assessors value all farm animals at only seven hundred and sixty dollars whereas the seventy-seven work boats are valued at twenty-five thousand and seventy dollars. Both figures are far below the actual market values, but they represent the disparity between farming and fishing. Yet within the last fifteen years there have been as many as eleven horses and thirty-four cows, not to mention a dozen steers and heifers. As recently as 1947, there were ten oxen.

At present there are no tractors on the island, though for a few years one was mentioned on the assessor's list. The closest approximations to tractors are various abortively contrived doodlebugs made of old automobiles, notable mainly for the great noise they are capable of. Since the eighteenth century oxen, aided later by horses, have done the work of the island. They were used in the fields, the woods, and alongshore to drag up boats for wintering. Within the past ten years wood has been largely replaced by kerosene as a cooking and heating fuel, so that there is no longer much need for animals to haul logs out of the woods. Trucks have taken over much of the hauling where roads permit, and also serve to pull the boats above high-water mark. Haying, logging, and plowing are done by the one ox and three horses still remaining.

At present several men operate small saws for "manufacturing" trap laths from hackmatack and spruce. In addition there are a couple of buzz saws for cutting stovewood and frame lumber for the traps. A considerable number of men still do some work in the woodlot during the winter, although not nearly so much as before the advent of oil, gas, and electricity. Occasionally someone has to replace the spilings of his wharf. The island's last working ox must be used for this purpose, as well as for swamping out cordwood. Fellow islanders who have tried to borrow him can do nothing with him, for his owner has spoiled him by talking human language instead of the customary whoa, giddap, gee, and haw. The owner explains in detail what he wishes the ox to do: "Now just back up a little mite and then turn to the right of that stump. You know damn well you can't get through to the left." The ox does as explained. If the load is heavy the ox sometimes stops and moos. The farmer doesn't yell, "Get up." Rather he expostulates, "I know it's heavy, but we don't want any back talk now. What we need is power." The power is supplied and as the beast strains along with log or wagon atow the farmer explains to the amazed onlooker: "He's just like a human being. When he moos like that it's just the same as you or me cursing. You've got to reason with him, that's all."

A completely equipped smithy once flourished on the island and is still prolonging a dilapidated existence. The blacksmith mended wheels, straightened axles, forged anchor chains, and made scallop dredges. He is said to have kept "maps" of the hoofs of all the horses on the island. A customer wanting a shoeing job done would telephone ahead and by the time he arrived the blacksmith would have the shoes ready for nailing on. The present owner of the shop still occasionally shoes his own horse there, but he does no public business. The shop is a relic. The bricks of the forge—in which are still the clinkers of long extinguished fires—are falling out of place. Hammers, tongs, chisels are mixed indiscriminately in heaps of waste iron. The long racks for holding shoes are depleted except for one or two worn and bent ones with twisted nails still in them. In the back of the shop, beneath sagging rafters, is the mighty "patented" sling for subduing recalcitrant animals. The gouged and grimy floor is flecked with sunlight that pierces the decaying roof.

Several farmers produce milk for local use. The usual way of pasturing the cows is to tie them out in the fields on chains. The effect is much the same as mooring lobster pots and may have been suggested by that practice. At present fences are almost nonexistent, though in the past when there was more livestock, wooden, wire, and stone fences were common. Haying is seldom completed till deep into September. Since no single farm has enough meadow to produce hay for more than a few animals, the farmers mow in various parts of the island. Some of them own the scattered fields on which they cut; others buy standing grass wherever they can. Summer people gladly give the hay from their fields so that the mowing will prune back the encroaching forests.

The farmers deliver their milk with a fisherman's disregard for the landsman's more servile notions of time. At any hour of day or night one may hear the clink of milk bottles on the back porch. No island milk is pasteurized, so that with the coming of electric refrigeration in the stores much pasteurized milk from the mainland has been sold at thirty cents a quart.

3

Some island soil is capable of intensive cultivation. Good oats and clover and timothy as "high as the horses' necks" can be grown if the land is given half a chance. One family has been particularly successful for many years in raising fabulous vegetables. The land is a lovely meadow on the northeast shore with views of the Mt. Desert hills and the open sea beyond the smaller islands. Generations of farmers have hauled up and spread tons of the rockweed washed up on the beach that forms one fringe of the meadow. The soil has become a fine, sandy loam.

The two staple crops of the present owner are cucumbers and potatoes, though it would be difficult to name a vegetable that he does not grow in marketable quantities. He plants potatoes in dead furrows with two regular furrows on either side. Most New Englanders, used to soil so shallow that a dead furrow usually lays bare the hardpan, will appreciate the kind of land that allows this sort of planting. These dead furrows contain loam deep enough to grow Kennebec potatoes whose vines flop three feet on either side of the row, and soft enough so that the farmer customarily digs the tubers, some of them two pounds in weight, with his bare fingers more or less as one gets eggs from under a broody hen. Small ones can be left to grow bigger.

Cucumbers are planted in small round plots off by themselves. The jumble of vines suggests that the seeds have been sown broadcast. Around the plots are short posts on which canvas is spread to keep off early frost. The yield is so plentiful that "you can bail the cucumbers out with a shovel." The demand on the island for this vegetable is great. Children like cucumbers almost as much as candy, and housewives are always eager for them.

Much of this produce the farmer sells at his farm to islanders who come in their cars to make their purchases. He encourages them to go into the garden and pick what they want for themselves. Sunday is the big day for this type of marketing. During the week the farmer hitches his horse to a dump cart and goes creaking across the island with a cargo of cucumbers and potatoes—which two he considers his "collateral"—and by way of specialties, green beans, peas, beets, or other delicacies that the season has to offer. The progress across the island is snail-like. Stops made to sell a cucumber develop into lengthy visits. Frequently it is long after dark before the equipage rumbles back past houses already in the hush of sleep.

This man is the supreme individualist of the island. He does things exactly as he wishes, which in general is in a way no one else has ever thought of. He has been written up in several periodicals and had his picture in at least one book. Post cards of his "double team"—the ox and his two-thousand-pound horse—hitched to a cutter and standing in the snow before the Baptist Church are for sale in island stores. He is quoted as saying that he considers one of his functions in life is supplying local color for the rusticators.

The method of fertilizing used by this family has had a curious and brilliant extension. For many years some of the most spectacular tuberous begonias in the world were raised on Swans Island. The grower was a Pennsylvanian who went there to recover his health fifty years ago. That first year he only stayed several weeks in one of the boarding houses. But he returned year after year, finally remaining for all but the midwinter months. He bought himself a house on Mackerel Cove on the north side of the island. To make it thoroughly weatherproof he encased it in an outer shell of granite blocks from the quarry then operating on the island. This "house inside a house" caused much consternation and finally, because of its practicality, a grudging approval. When the owner equipped it with an electric plant, dug an artesian well, and installed plumbing, emulation set in among the wealthier islanders. He extended a wharf off his beach with a landing float at the end. As a hobby, this same man began to experiment with tuberous begonias—those geniuses of the potato family which can imitate and surpass the most aristocratic flowers in beauty. On Swans Island they came dazzlingly into bloom. The father of the truck gardener just described told the begonia enthusiast how to prepare island soil with rockweed. From then, he said, everything he tried "panned out." He planted evergreen hedges to keep the incessant winds off the beds, in each of which he had up to twenty-five hundred tubers. He erected a greenhouse and a potting shed and hired an English gardener, who stayed on the premises the year round. From about 1920 to 1945 he did a booming business. Some years he made up to ten thousand dollars, selling from twenty to forty thousand tubers all over the nation and abroad. Millionaires on the summer resort islands of North Haven and Mt. Desert became interested and visited Swans Island in their yachts. Orders were placed for flowers for fashionable weddings. Tubers sold for sums up to five dollars apiece. It is said that Lady Astor, who saw some of the flowers while she was visiting in North Haven, ordered some roots to be taken back to England, though England supposedly grew the finest begonias in the world.

No horticulturist who saw these begonia beds hesitated a moment in admitting they surpassed anything he had ever imagined. They were a floral wonder, growing beside the gray sterile ledges that everywhere break through the soil. The tubers reproduced every flower—roses, camellias, daisies, pansies, orchids—in even greater size and more brilliant and varied bloom than the originals. There was fantasy in it—nature gone berserk, plunging through the barriers of her own laws in a wild explosion of color. Those who came to wonder had one burning question: "How is it done?" But there was no secret. The owner could only repeat what he'd said a thousand times. He had followed a local tradition in applying seaweed to the island earth that lay so shallow and unpromising among the granite outcroppings.

Swans Island begonias no longer grow in their original soil, but their descendants are scattered in splashes of color among the gardens of the world. The old beds behind the windbreaks sprout mammoth stands of weeds. The greenhouse is slowly dropping its panes. The planks of the verandah of the dwelling are rotting and the wharf has tumbled into the sea. Ill health and old age have driven the owner out of what he aptly described as "one of the nicest businesses imaginable."

Writers describing the Maine coast in the last century sometimes mention the flowers growing around the squat little houses. With the exception of the ubiquitous lilacs and an occasional residence, this is not the case on Swans Island. Yet the islanders have a strong appreciation of beauty. They admire the whitecapped water beneath the cottony sweep of clouds on a northwest day, the incredible vermilion of sea and sky as the sun sets behind the sombre spruce-covered islets, the nearness of the distant hills to the north in the crystal days of autumn. The wonder is that they don't plant more flowers, which they know will bloom superbly, to furnish color to their own scrap of land.

The decline of gardening is perhaps part of the islanders' more and more specialized lives as suppliers of lobsters for the nation's tables. The island becomes less a home to be made as beautiful and resourceful as possible, and more a place where families are stationed to carry on a specific job.

The case of Criehaven, the outermost of all Maine islands, is a warning example. When lobster prices are way down, Criehaven (on eighteenth-century British admiralty charts named Ragged Arse Island) is deserted. The inhabitants shut their houses and go ashore to earn whatever living may come their way. When lobster prices rise again, the inhabitants return. Now most of them live there only through the clement months when the weather permits reasonably continuous fishing on these turbulent outer waters. The rest of the year they live on the mainland or one of the larger islands where they often own second homes and their children go to school. This loveliest of Maine islands, described in Elizabeth Ogilvie's novels, has become little more than a base of operations for lobstering.

The present tendency is one of pulling up roots buried for generations. Families are scattered from Eastport to San Diego. On some islands, like Isle au Haut whose permanent population has shrunk from five hundred to fifty within this century, the uprooting is far advanced. On Swans Island—most conservative, some say most reactionary, among the islands—the roots are more firmly anchored, but still they are beginning to tear away. The passing of farming and gardening is the first and most serious sign of eventual desertion. The soil is one of the strongest ties that hold people to a locality. So long as that hold continues, people can share the attachment that a nineteenth-century village writer of New England expressed for the place of her birth: "It is strange that the spot of earth where we were born should make such a difference to us." But this feeling for place, which is what we mean when we speak of roots, vanishes very quickly when the attachment to the soil is severed. Not only the fields and pastures of the islands are being smothered in underbrush. The memories and emotions that hold people to the home of their forefathers are slowly suffocating too.

4

On Swans Island a surprising number of people make a living, or add to what they already have, by opening a store. The would-be storekeeper moves the furniture out of the back room of his house, gets a Coca-Cola or Moxie sign, orders a stock of cereals, tobacco, beans, sugar, molasses, flour, candy bars, some patent cathartics and cold remedies, pencils and paper to catch the school trade, and a few knick-knacks such as jack-knives, can openers, and fly swatters. Sometimes he purchases his stock from a retail chain store on the mainland and marks the prices up, but usually he opens an account with a regular wholesaler. He installs a soft-drink cooler for his tonics, an electric refrigerator so that he can carry ice cream and a line of frozen meats and vegetables.

At present there are six or seven such stores. Two of them are combined with post offices, a most favorable arrangement since the proprietor can be sure that most of the people in his village will be exposed to the lure of his merchandise when they come in for their mail. With the advent of electricity one has set up a soda bar and a table at which customers can linger over their soft drinks and ice cream. Remarkably, with a population of only four hundred and fifty in winter and five hundred for a month or so in summer, these six stores not only manage to survive but actually are able to expand. One reason that they persist is that their owners have other resources. Two enjoy subsidies from the United States Government in the form of post offices. Others are run by one member of a household, perhaps a wife or daughter, with no expectation of supplying more than a part of the family income. Several shopkeepers do some lobstering or fishing. Only two rely entirely on their stores for a living.

Another reason for the survival of these enterprises is the absence of overhead. Located in private homes anywhere on the island they are not taxed like business property on the mainland, though there is a town property tax on all "stocks in trade," that is, on the inventories of the various stores. Of course, there is no rent to pay. Several of the storekeepers hire a part-time helper, but the wages are low. Prices are higher than in small neighborhood stores on the mainland and much higher than in chain stores. Milk, for example, costs twenty-nine cents or thirty cents a quart, margarine forty cents a pound, and ice cream thirty-seven cents a pint. Yet considering the freight charges for getting these goods across the water, the prices are not so exorbitant. There is plenty of competition to keep a ceiling on them.

Most of the island tradespeople show a great deal of initiative, even if it is on an almost ludicrously small scale. The ice-cream parlor and a newsstand are examples. Wherever there is the barest chance of attracting a little trade, the proprietor will not hesitate to have a try at it. One store sends a grocery truck around the island twice a week. With barrels and cartons tied on its side and the interior shelves stocked to the roof, the vehicle carries as complete a line of groceries as any of the shops themselves.

This same store also attempted one of the most daring business ventures in many a year. Buying a mobile hot-dog wagon, the proprietors set it up at the chief intersection on the island. The hope was that they would sell grilled frankfurts and soft drinks to the Sunday drivers. On the mainland it would have worked, but on the island's 12.85 miles of roads it is impossible to stray more than four or five miles from one's home by auto—hardly far enough to work up a thirst or appetite. But for part of a summer the wagon stood hopefully in the center of the intersection. It enjoyed one spurt of business when the state road workers came for their annual repair jobs. Shortly after they departed, it was parked behind the owner's store, where it still stands.

Few islanders trade exclusively with one store, partly because the stock of each is so limited that one couldn't fill all his needs from it. If one visits enough stores and his tastes are simple, he will usually find what he wants. People who have a grudge against a certain proprietor will avoid his place but will spread their trade among the others. Such grudges usually do not arise from a business deal but from some other controversy, like the boat dispute, in which several storekeepers made enemies by taking a violently partisan stand. Sometimes the grudge is more personal, the aftermath of a dispute over a well, a boundary, or a right of way. But unless influenced by some such animosity the islander regards it almost as a duty to shop around. He feels mildly responsible for keeping all the stores going. For some a more selfish reason for spreading their trade is that a steady customer at several stores has more places at which he can get credit when the need arises.

Most islanders do some shopping on the mainland. Prices there are usually much cheaper, so that an occasional shopping trip in one's own boat can be a real economy. The storekeepers complain that this off-the-island trading is so extensive that it seriously menaces their business. They argue that they are doing the island a service by maintaining stores against staggering difficulties and that the islanders should appreciate these efforts by shopping at home. A fisherman who happens to be in Southwest Harbor, Stonington, Bar Harbor, or Ellsworth—the four most popular spots for mainland shopping—shouldn't stock up at the A & P, even though on some of his purchases he may save up to 50 percent. Local loyalty and gratitude should send him home empty-handed to spend his money among his benefactors. However, some business comes to the island from yachts cruising the coast who slip in to restock.

Many goods, clothes in particular, can no longer be bought on the island. In the past there were three general stores, which stocked a phenomenal variety of goods and served as gathering places for islanders with some idle time to chew the rag. But two closed with the deaths of their owners, and one has gradually sold off its stock. Now if the islanders are too busy to take a trip to the main, their only recourse is the mail order houses, usually Sears Roebuck, or Montgomery Ward.

And inadequate space for cracker barrels discourages the use of the new stores as social centers.

The most profitable island business has always been the buying of the fishermen's catches, combined with the selling of marine supplies. The buyers of former days often got business from off the island. Boats would come in to sell their fish and stock up before making another trip. Another source of business was outfitting local boats sailing to the Grand Banks or the Gulf of St. Lawrence. The building in which one of the first of these stores was located still stands. It is a long wooden, peak-roofed structure suggestive of a warehouse, near the site of some distintegrated wharves where schooners used to tie up. It had to be capacious to hold the sails, barrels, and coils of ropes needed for extended fishing trips. When island schooners no longer sailed on distant voyages, the building was variously used as a boatwright's shop, a general store, a café (pronounced *caif*), and finally as a summer cottage.

Today there are no fish buyers, but the lobster buyers operate in much the same way. They purchase the catches and sell gasoline and equipment such as warp, paint, trap heads, and bait pockets. If a fisherman loses his gear in a storm or has a series of poor catches, the buyer will extend credit, with the tacit understanding that the fisherman will sell to him when his luck changes. Three buyers do business in Burnt Coat Harbor, each one reselling to a different company in Rockland. Until recently a fourth did a small business from a scow in Mackerel Cove. On every pound of lobster that the buyer handles he makes two cents. One of his main functions is to take the fishermen's catches off their hands and store them in "cars," huge wooden cages afloat in the water, until the smacks make their weekly calls to carry the accumulations to the distributors. These local buyers are useful, and probably necessary. Usually they are the leading businessmen in the community. If they are at all successful they build up a group of loyal fishermen who do business with them year after year.

In any Maine harbor the lobster-buying wharf or scow is a center of activity. Boats are continuously chugging in to unload a catch or take on fuel. The proprietor and his helper scurry about with tubs of lobsters and long-handled nets. Fishermen shout their interminable jests at one another in voices stentorian enough to rise above the grumble of a half dozen motors. A cruising sail yacht, graceful and aloof among the ungainly workboats, glides in to purchase gasoline for its auxiliary motor or to tie up for a little shopping at a nearby store.

5

In the early days many islanders made money by gathering cobbles on the beaches. When they had loaded them from dories onto schooners just offshore, they sailed to Boston Harbor and landed them at the Stone Wharf in Charleston to be used for paving streets. When cobbles went out of fashion, the islanders started splitting off the primordial granite. By the end of the last century, almost every island from Wiscasset to Mt. Desert had its quarry. On the tiniest islets today one can see the carved ledges and loading wharf, durable as the shore itself, butting cliff-like into the water.

The best known quarries were at Deer Isle, where stone was cut for many noted buildings, among them the Cathedral of St. John the Divine in New York City. The advantages of these island and shore quarries was that the blocks could be loaded on barges on the spot and towed by water to their destination. Sometimes they would need no rehandling, if they were to be used in a bridge, a breakwater, or a dock. The leavings from the quarries often were hauled away to make wharves on the nearby coast.

Many a New York street is paved with stones gouged from the shore of Burnt Coat Harbor. When fishing was poor, islanders with their hammers and boxes of chisels and wedges hacked away at the escarpment, forcing it deeper and deeper into the hillside. At other times the fishermen were replaced by Swedes, as many as two hundred of them, who bunked in lodging houses around the village and left when work was slack. On the shore below the quarry was a wharf for loading the blocks. A railroad ran from the quarry top to the dock. It was an industry that the island was proud of. Twenty years ago the enterprise was abandoned. Demoralization had set in when one of the owners was atomized in a dynamite accident. But more serious, paving stones were going the way of cobbles as surfacing for streets. Automobile traffic demanded something smoother, like asphalt or concrete. One morning the ring of hammers no longer awakened the echoes around the harbor. Today the deserted cut frowns above the east side of the harbor. To the hamlet of Minturn just beneath, its barren heaps of waste and its stern walls give an air of northern desolation. The old firebox and rusty boiler, cold as the stone, lie prostrate among the chips and leavings. Wild orchid hyacinths flourish in the bleakness. The pit is flooded with deep, clear spring water where the gulls and the village boys swim. Up there under the sun the fresh water warms to a comfortable temperature. Trout placed in it by the state keep it pure. Below on the great wharf where the barges used to tie up, the islanders gather twice daily at the departure and arrival of the mailboat.

The islanders regret the loss of the quarry. It helped many a fisherman who was down in his luck. But it is now only a memory in the minds of the older men, a few of whom have kept their stonecutter's kits for occasional use in splitting and shaping a doorstep or mending a foundation.

Lumbering has fared no better than quarrying. In the early

days many boards were sawed at Swan's tidal mill; later the second growth of island wood was cut and shipped in schooners to Rockland for use in the limekilns there. After the island was organized as a plantation, the wood on the common land —*i. e.*, the land unoccupied and unclaimed by any settler— was available free of charge for any islander who wished to cut it for the kilns. One man walked clear across the island, six miles each way, to cut his daily cord, which he could sell for eighty cents to one of the lime companies.

But lumber has not been shipped away in many years. Once in a while stands of spruce are sold on the stump to pulp mills, but difficulties of shipping discourage the exploitation of the extensive pulp forests. A cord of pulpwood brings only eight dollars at the shore ready for loading into barges. Since one cannot hire a man to cut a cord of wood for less than six dollars, the margin of profit is prohibitively small. The trouble is attributable to the cheap pulp produced by the Maritime Provinces which, as an islander thinks, have all but "blotted out the State of Maine's industries."

The island has an old tradition of craftsmanship—mainly in ship-building—which still lingers among the two or three remaining carpenters and painters. Some of the earliest settlers were boatwrights who came to the island to take advantage of the oak timber that was so plentiful there. Island oak has long been considered the finest material for boats, for trees that have grown in salt air are believed to be more resistant to salt water. Swans Island ships had a fine reputation and one, the *Henry M. Johnson*, was ordered from as far away as Newark, New Jersey. Another, the *Arcadie*, was still thoroughly seaworthy after sixty years of use. Many an islander was not only a farmer, a fisherman, and a mariner, but also— to top off his versatility—an expert boat-builder as well. When shipping declined throughout Maine, small boats were still built on the island. The craft is not quite dead even today.

One boatbuilder, whom the islanders designate as "the best damned boatbuilder in the state," still does painstaking and highly skilled work entirely with hand tools.

One or two other men are competent house carpenters but in no way exceptional. Others are housepainters. All of them work with great deliberation and care, but their chief attraction to summer residents is that they charge one dollar and twenty-five cents an hour. Much of their work is for summer people, a lot of it in simple repair jobs or in reconditioning an old house, an undertaking they all enjoy. In renovating an old house a workman calls to mind many old associations. Undoubtedly an islander built it, perhaps an ancestor or relative of the present artisan.

Many of the trades and professions found in even the smallest mainland towns are not represented on the island. There is no regular barber though one or two fishermen do a little haircutting as a sideline. There are no plumbers, cobblers, electricians; in all these trades only the services of an amateur are available to those adept in the art of cajolery. With electrification of the island, the need for skilled labor became more apparent than ever. Only one man knows how to care for the diesel motors of the electric plant. Yet opportunities for many other technicians knock loudly. For example, with all the boats and machinery there is no one capable of welding. One characteristic of the Maine fisherman, noted years ago in a United States Fish Commission report, is his unwillingness to work for someone else, even though he happens to be idle and in need of money. He may be able to act as his own carpenter, mechanic, painter, farmer, or electrician, but he is reluctant to give his neighbor the benefit of his versatility. The odd job is definitely not to his liking. Thus the cry of islanders and summer people alike: "You can't get anyone to do anything for you."

6

Where every other means of livelihood fails in rural New England townships, the roads are an enduring source of income. The number of Swans Island families saved from the humiliation of "going on the town" by a timely offer of a road job would be a sizable percentage of the population. Maintenance of the roads is a sort of perpetual WPA. Rather than putting a man on the poor rolls, the town fathers, if they know their duty, will set him to work with shovel and pick or brush hook keeping the public thoroughfares in order. The road commissioner, elected annually to his position, is an important public officer. He and one or two helpers make almost a full-time job of it. It is their profession. They understand the principles of road mending and the use of the necessary machinery. A dozen or so others work under their direction when some big project is under way—a heavy snowfall to be removed, trimming away the brush and grass along the edges, the regular scrapings and oilings.

Some of the money paid in wages is tax money raised from the town. But much of it is contributed by the state, so that road work is a way of bringing in outside money, and the shopkeepers benefit from it as well as the laborers. In the year from February 5, 1953, to February 7, 1954, the town received $4,461.84 from the state for the care of its 12.85 miles of road. The town itself added only $2,266 for the same purposes. The town officials make a point of using the state allotments down to the last penny, if possible, so that the next year's sum won't be smaller. In mild winters considerable perseverance is needed in using the snow-removal fund of sixty-five dollars per mile. Every half-inch fall has to be taken advantage of. One islander wrily remarked when he saw the plow go by after a dusting of snow had fallen, "Well, anyway they're tearing up the hard-top so that they can spend next summer putting it back on again."

All of the money for road maintenance does not, of course, find its way back into the town. Equipment like power shovels and graders and men to run them have to be brought from the mainland, as does most of the material such as oil and cement. Yet of the combined town and state appropriations of $6,194.84, the sum of $2,590.29 was spent for town labor, $985.01 for the hire of trucks owned in or by the town, $221.25 to a local woman for boarding state road men, and $455.47 to island stores for gasoline and "supplies." The income to the town from these expenditures would be $1,986.02 more than the town appropriated for roads on its own budget, or a net gain of $154.55 on each mile.

The roads that receive the most state aid are those that connect the three villages on the island. Beyond the villages the hard-top quickly dwindles out into "end roads," which are supported primarily by town funds and thus are in dismal disrepair. Those who live on the end roads, particularly on their farther reaches just before they become woodpaths, are often at war with the selectmen. They are sometimes deprived of school-bus service and often of the use of their own car. Small lakes of water a foot or so deep across one of these roads prompted a man who lived at the end of it to telephone the selectmen to send out "either a few loads of gravel or a punt so that he could get in to town to do some trading."

But the hard-surfaced, state-aid roads compare with the best on the mainland. They are broad, banked for high speeds on the curves, posted with traffic signs, and protected with white-painted guard rails. Some townspeople question the need of such speedways on an island where the villages are about three miles apart and the two most widely separated houses are not over six miles apart. On Matinicus Island, which has no roads other than rutted paths through the fields, it is impossible to travel faster than ten miles an hour.

Good roads bring all the hazards of automobiling found on the mainland. Teen-agers are the worst but not the only offenders, and speeding, of course, is the prevalent offense. On the two or three straight stretches, the island drivers press the accelerator to the floor. "Stop" and "Caution" signs mean even less to them than to mainlanders. The lives of children are menaced and those of cats and dogs frequently crushed out. Bad accidents, including at least one that resulted in a fatality, have occurred. Driving under the influence of alcohol is often reported. Cars race into curves so fast that they sometimes plunge through the guard rails into the woods.

Everyone knows the offenders and complains of them. In the worst cases warnings are given, but their effects are transitory. The intimacy of island life restrains anyone from prosecution. It is easier and friendlier just to let things slide. Arrests and fines would only bring hard feelings between neighbors and might also cause family rows as well. All responsible islanders do their best to protect the community from the corrosive feuds which, once started in the closely knit life lived there, are almost impossible to smooth over.

The State Police seldom get to Swans Island, and when they do they accomplish little. Their arrival is known to everyone the moment the boat docks. Unregistered cars are kept off the road; the town suddenly possesses a population of cautious drivers. When the police leave—"those striped devils," as one islander called them alluding to their uniforms—the people can breathe easily again. They can be sure there will be no more intrusions for another year.

One incensed citizen who regards automobiles as an invention of Satan said there were two cars for every person on the island. This is a splenetic exaggeration, but there is a total of at least one hundred. Formerly most of them were incredible jalopies, fragments of many cars bolted and wired together to form a contraption that could barely thunder and clank its way across the island. Only a few persons, like the lobster buyers, could afford cars in reasonably good condition. But with improved roads and higher lobster prices, the most impossible of the jalopies were replaced by presentable secondhand and even new vehicles. And with better cars came the higher speeds.

Visitors coming to the island for the first time are usually overcome by the advanced state of automobiling they find. Landing on the quarry wharf they are first struck with the simplicity and quiet charm of the scene—the houses huddled along the shore among the spruces, the church spire etched against the northern horizon, the quiet water of the harbor with its white workboats lolling at their moorings. Then in the distance is a hum which like an oncoming tornado crescendos quickly to a roar. A truck skids around a bend and screeches to a stop, its front wheel a foot from the outer edge of the wharf. A gangling young fisherman with rubber boots aflap slides from the cab and grins fatuously. For the visitor the spell is broken. During a five-hundred-mile drive, just ended, on the main highways of four states he has witnessed nothing so terrifying. Sadly he realizes that there is no place to hide; that he was foolish to expect anything different. This remote, spruce-clad island where wistful summer people fondly hope they can escape the more unpleasant aspects of the civilization with which they are beset all the rest of the year is still a part of America and the islander is thoroughly American. He glories in owning a high-powered vehicle to convey him a few miles at breakneck speed, with fearful jeopardy to his own and others' lives. He has been told over his radio, and recently on his television screen, that this is what he most ardently desires. He has come to believe that he does desire it, despite the canniness of his Yankee mind, and now he has it.

But the reckless drivers are still in the minority on the island. The average islander sticks to his aged but usable second-hand car, which he drives with consideration for himself and his neighbors. Several have bought small English-type cars which burn little gas and which can be brought to the island on the mailboat. Others have jeeps, which are most practical for the poor roads off the main highway. One or two use motor bicycles to the exclusion of automobiles. And a number find that they need no form of land conveyance. All three villages are built on the shore and are accessible by boat. For all families the boat is the first need anyway. The automobile is a luxury.

7

Isolation and Dependency

SWANS ISLAND HAS ALWAYS BEEN NEAR THE MAIN HIGHWAYS OF history, perhaps a short way down a bypath, but always within hailing distance. In colonial times it was in the midst of territory disputed by three nations. In the summer of 1613 an Indian or white fisherman on the north shore would have heard reverberating from the slopes of Mt. Desert some of the earliest shots of the interminable wars between the French and English over the control of North America. In that year the Jesuits established their first mission in New France on the shore of the fiord-like Somes Sound. Within weeks of their arrival a British ship under Argall, dispatched in frantic haste from Jamestown, sailed up among the islands, entered the Sound with the aid of Indian guides, and opened fire on the Frenchmen. It was not much of a fight, but one missionary was killed and the others taken prisoners. Later at Castine, a few miles to the westward, a three-cornered rivalry among the Acadian French, the English from Plymouth Colony, and the Dutch from New Amsterdam sent periodic grumbles of cannon down Eggemoggin Reach. At the end of the seventeenth century a French census showed a scattering of Acadian families with their animals and well-furnished homes all along this part of the coast. At nearby Gott's Island (then called Petit Plaisants) were two families. Cadillac and his wife were on Mt. Desert, and Baron Castine was at Penobscot. Farther westward families were reported on Matinicus, twenty miles offshore, and on Metinic. But the noises of war were never long silent among the islands from century to century, through the Revolution and the War of 1812, down to the international conflagrations of our own times.

Wars to the islanders are like storms in some distant latitude of the sea. Their centers may be thousands of miles away, but their remote tumult is heard in the surf that swells out of the ocean and mutters along the shore. Sometimes the disturbance sweeps in closer and the surf thunders more ominously, as in our recent wars. World War II brought enemy submarines even into the shoal waters of nearby fishing banks where they would surface and at gun-point relieve a solitary fisherman of his catch. Depth charges trembled above the clatter of the surf. In island homes pictures fell from the walls and window panes were cracked. Squadrons of planes swept back and forth along the coast and observation blimps hovered like summer clouds above the bay. One blimp was shot down, apparently by a submarine near Mt. Desert Rock.

In one secluded summer cottage on Swans Island, from which a view of open water reached out to the horizon, evidence was found of the winter work of spies—a window shutter removed for the shining of lights out to sea, empty food cans, and burned candle stubs. Nearby on the mainland enemy agents came ashore in rubber boats from submarines and made their way inland through the snow.

One summer morning islanders on the northern shore awoke to see the Presidential yacht at anchor a mile or two up the bay. Eastward along the horizon an unbroken line of battleships blended into the gray of the sea. Speedboats sizzled through the waves from Mt. Desert to the yacht and back again and planes circled restlessly overhead. Within a few hours the navies of two nations vanished as abruptly, as inevitably, as they came. Later the only mildly surprised islanders read in the newspapers of the signing of the Atlantic

Pact in the waters off Mt. Desert. Thus the streams of war flow in and out among the islands, though they haven't flooded over the shores nor lingered long in their midst. Clasped as it is in the sea, Swans Island is enveloped in history. It affords no place to hide from human destiny.

Men from the island have served in all the nation's wars. During World War I some were taken into the Navy, but a surprising number were in the Army. In World War II there was one casualty from each of the three villages. The Civil War made the greatest impact on the town, but the men showed a marked disinclination to fight. In 1863 the plantation voted to "hire" twenty-one hundred dollars to pay for volunteers; the next year it voted seventy-six hundred dollars to reimburse the men who had put substitutes in the Army under the call of 1864. At the same time it raised eight hundred dollars for every man who enlisted. In 1865 the plantation meeting "heard the report of the men who had put substitutes into the Army" and "voted to pay them 50 percent of what they had paid out, and exempt them from further taxation." In the Presidential election of 1860 this community, which is now solidly Republican, had given the Democratic electors twenty votes and the Republican nineteen.

The proximity of the island to Canada has made it a convenient spot for smugglers from its first settlement down through the prohibition era, during which many a boatload of whiskey is said to have found its way into caches beneath the water alongshore. Prior to the War of 1812 the cellar of the "Big House" (Swan's mansion) served as a depot for contraband goods to be shipped into British territory. But the Revolutionary War veteran known as "King David" Smith reported these activities to the nearest customs officer and helped him load the goods on a boat for shipment to Deer Isle, where it was confiscated. For his pains "King David" was later assaulted by two smugglers, but true to his royal nickname he thrashed them single-handed. When brought before a magistrate by his victims, Smith was acquitted on the grounds of self-defense. Opportunely, the magistrate was also the customs officer whom the defendant had aided in making the seizure.

2

For a hundred years the island had no regular boat service to and from the mainland. Anyone wishing to go "ashore" had to sail his own boat or hire someone else to take him. Letters were carried over by anyone who happened to be on the mainland. In 1844 the first post office was established, and with it a weekly crossing to Brooklin to pick up the mail. Later, trips were made twice weekly and eventually the government established a daily delivery. Until 1894 the mails had always been conveyed either to Mt. Desert or Brooklin, but in that year the contract was let to a local steamboat company that had inaugurated service between Swans Island and Rockland thirty miles to the west.

A mainlander, particularly in the automobile age, can have no idea how important to an island is its "Boat." The "Boat" is not only the mailboat, but the freight and passenger boat, the one steady, dependable tie with the "main." Her daily arrival is an event, no matter whether she is a steamer of over one hundred feet in length or a converted sardine smack of a scant fifty feet. Men, women, and children gather at the wharf at boat time to welcome relatives or friends, to pick up freight, or simply to be there. From the hill above the harbor the color and activity on the wharf are accentuated by the sombre year-round hues of the setting—the water, the granite ledges, the houses all a monotonous gray and white, and the shores draped in the dark, funereal green of the spruces. The insularity of the town is brought into relief, as is the solidarity of the people thrown here together on this sea-hugged scrap of land. This is their moment of recognition of their connection with the rest of their kind.

The whistle of the boat, always sounded as she rounds the harbor head, becomes a part of the lives of the people. It is a milestone in the progress of the day, a cause of satisfaction if it is heard at the scheduled time but of alarm if it is late. Those who live across the island out of hearing of the whistle wait for the single long "boat ring" that central sends out on the telephone lines. It is not only a warning that the mail will soon be ready in the post offices. It is also an assurance that the island is still in touch with the mainland, is for one more day at least not forgotten. So attached do the islanders and boat crews become to the whistles of their boats that with the scrapping of a favorite craft her whistle is often installed on her successor. No two whistles are alike. Some are chimed, some are sirens, and each one has its subtle differences.

Regular and dependable service to the Maine islands came only with the steamboat, at first great awkward sidewheelers and later vessels driven with screw propellers. The terminus of most of the eastern island routes was Rockland. Altogether over forty steamboats have at one time or another been on these runs, not to mention the many coastwise and Penobscot River boats either terminating or making stops at Rockland. The nearby islands like Vinal Haven were the first to get this service. After 1857 Swans Islanders became accustomed to seeing the Bar Harbor and Blue Hill sidewheelers splashing down through Eggemoggin Reach or Deer Island Thorofare and across Blue Hill Bay to Mt. Desert Island. The passing of these boats, some of them large and luxurious, became a part of the passing day. Ships that had been long on the route, like the *J. T. Morse*, seemed like old friends even though they didn't stop at the island.

But Swans Island was not long overlooked. Steamboating was booming on the Maine coast, particularly east of the Penobscot where there were no railroads along the shore till almost the twentieth century. With the great prosperity of the mackerel fishery in the '70's and '80's, steamboat connections were established between the island and Rockland For a time steamboats stopped on both sides of the island, but usually the only port of call was Old Harbor, where a wharf and a landing slip and warehouse were built. Though boat service was discontinued in 1941, many islanders still consider Rockland their mainland base, a place in which to shop and to have a good time, and the home of many relatives and friends.

Schooners and squareriggers, snowy canvas set for the far reaches of the world, may be much more romantic, but there is drama in the story of the Maine steamboats. Winter and summer, in blizzards and in fog so thick you couldn't see the bow from the pilot house, the steamers groped their way among countless ledges and islands and along the narrow, sinuous "thorofares" where the tidal current races like rivers in a freshet. This is the area that Champlain chose to enter just once and then "with sounding lead in hand," and which as late as 1765 was marked "No Passage" on official charts. As the steamboats crossed the open spaces between the larger islands, mountainous seas would sometimes roll in from the ocean, and at the far side was always some strait or harbor mouth where a twenty-foot deviation from the course might mean disaster.

During the sixty years of steam service, three or four ships and two captains occupied a special spot in the heart of all Swans Islanders. The first vessel to stop was the *May Field*. She was built in 1875 for the Barbours, a family of Swans Island. For a few years Captain Alvah Barbour, master and

Colonel James Swan
(*From the portrait by Gilbert Stuart—Courtesy Museum of Fine Arts, Boston*)

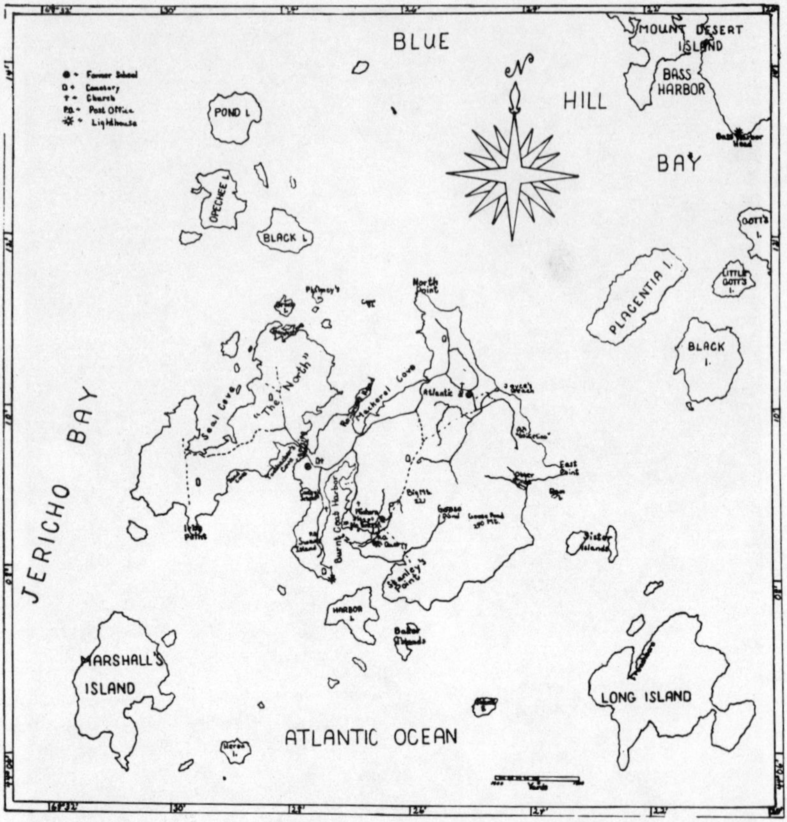

Map of Swans Island and surrounding area.

Fish house and wharf at Atlantic. (*Courtesy Arlen Birr*)

Burnt Coat Harbor. (*Courtesy Arlen Birr*)

Sardine boats. (*Courtesy Arlen Birr*)

The lobster fleet at Minturn.

Swans Island Light at Hocomock Head. (*Courtesy Arlen Birr*)

Oceanside at mouth of Blue Hill Bay.

Fish wharf at low tide at Old Harbor.

Productive lobster water in Toothaker Cove. (*Courtesy Arlen Birr*)

View up Blue Hill Bay. (*All pictures on this page courtesy Arlen Birr*)

View up Mackerel Cove toward Carrying Place.

East shore of Mackerel Cove.

Thunder Hole on east shore of island.

Rocks at Hocomock Head.

Low tide at Blue Hill Bay.

The same spot at high tide.

Joyce's Beach viewed from above. (*Courtesy Harry Baden*)

Joyce's Beach at sea level. (*Courtesy Arlen Birr*)

A typical fisherman's cottage.

An example of high-pitched roof favored by Swans Island carpenters.

Methodist Church in Swans Island Village.

The *Sunbeam*, the Mission boat that serves Maine islands.

Mowing hay on a seaside meadow. (*Courtesy Clyde Torrey*)

The oldest house on the island.

A farmer with an oddly paired team. (*Courtesy Douglas Dickie*)

Barn and pasture on Swans Island.

Swans Island mailboat, the *Sea Wind*, arriving at Bass Harbor. (Courtesy *Harry Baden*)

Workboat and wharves at Minturn. (Courtesy *J. T. Phinney*)

A commemorative stone in the graveyard at Atlantic. (*Courtesy Harry Baden*)

Deserted farmhouse in interior of Swans Island.

part owner, ran a route with the *May Field* from Rockland to Somesville on Mt. Desert Island. One of the stops was in Mackerel Cove on the north side, the section known as Atlantic. The wharf has long since disappeared, but the older inhabitants still point out its site, which is associated with the Golden Age when two steamers touched at the island.

Another early comer was the *Pantagoet*, originally built for service in the Great Lakes. She grossed three hundred and thirty-two tons and was one hundred and twenty-eight feet long. After being taken from the Swans Island run in the late '90's and put on a route from Eastport and Rockland to New York, she met the violent end that was the lot of so many island boats. She went down in the great gale of November, 1898, the same in which the sidewheeler *Portland* was rammed by a schooner and sunk off Cape Cod with a loss of one hundred and seventy-six lives. Unlike the *Portland*, no trace of the *Pantagoet* or her crew of sixteen was ever found. She was last seen off Highland Light, Cape Cod, creeping northward into the darkening storm. The old wooden "laker," not a complimentary epithet among salt-water sailors, was not built for Atlantic hurricanes. Before her last voyage a canny Swans Islander who was a member of her crew threw over his job, remarking that "the stick ain't worth the candle." No Swans Islanders were aboard when she went down.

The hundred-foot *Vinal Haven*, one of the most admired steamers, served Swans Island from 1898 to 1920. Her master, too, was Alvah Barbour, who year after year guided her back and forth with no mishaps other than a fire that temporarily sunk her at her Rockland wharf. It is a tribute to Captain Barbour that this boat which he commanded during most of her forty-six years died a natural death, somewhat a rarity among Maine coast steamboats. Those that didn't go to the bottom or burn up were often retired to the calmer waters of the Hudson, where after a few years on the Albany lines they were laid up somewhere to rot. The decaying bulk of many a

Maine boat can be found between Albany and Bear Mountain.

Captain Barbour was one of the famous skippers in Maine coastwise shipping, a man with the skill of one of Mark Twain's Mississippi pilot heroes and working in waters fully as difficult and much more dangerous than the Mississippi. Around the turn of the century, when Captain Barbour was at his prime, several rival companies were operating in the Penobscot region. Boats of these firms vied with one another in comfort, regularity, and, above all, speed of service. Boatmen set great store by speed. Far back into sailing days this was the coveted quality. Down to the present many a race, impromptu or planned, is run between workboats to settle controversies as to the superiority of certain designs and engines. Thus the one quality in a steamboat most likely to impress the island fishermen was speed, and other things being equal, the fastest boat would get the most customers. Steamboat "wars" of great bitterness broke out, especially on the lucrative run from Rockland to Vinal Haven. Wagers were laid on the competitors. Acrimonious arguments developed and doggerel was published in the local newspapers extolling the merits of each contestant and belittling her rival. Shouting throngs would await the boats at each end of the runs across the Bay. The *Vinal Haven* before Captain Barbour became her skipper had been a major contender in these races.

The competition called for skippers who could get the most out of their vessels without jeopardizing them or their passengers. The "wars" were confined mainly to the western part of Penobscot Bay and Captain Barbour was apparently not in the thick of them. Yet he was infected with their spirit. The Swans Island boats, which he piloted, made stops at North Haven and Stonington, in direct competition with other lines serving these points. One day on the western end of the run he was outstripped by a steamer called the *Sylvia*, which ran between Rockland and North Haven. Captain Barbour unfortunately was at the wheel of the slow-paced *Emmeline,* which once had figured prominently in the "wars" but had long since been outclassed. As the *Sylvia* glided past, her passengers raised a cry of jubilation. Captain Barbour in fury roared back: "You couldn't do that if I had the *Vinal!*"

The *Vinal Haven,* nicknamed "Old Faithful," made a record by stacking up forty-six years of service. But after 1920 she was taken off the Swans Island route and put back on the shorter one to Vinal Haven. Her place was filled by a contemporary and former rival, the *Gov. Bodwell,* whose master, Ross Kent, was another Swans Islander. The *Bodwell's* luck was about as bad as the *Vinal's* was good. On January 27, 1924, she struck on Spindle Ledge in the narrow western passage into Burnt Coat Harbor, barely a stone's throw from the lighthouse. It was one of the coldest days in the island's history. The temperature was sixteen below zero, a violent gale was lashing up spray and driving vapor in impenetrable clouds, a nearby bell buoy, vital in making a blind entrance into the harbor, had frozen silent, and at the very moment the ship was groping for the channel a heavy snow squall closed down.

Boats have made the harbor in equally foul weather. A Captain Ingraham took his steamer out of Bar Harbor in a blizzard so violent that destruction threatened him in that exposed port. He steamed southward through ice-bound Frenchman's Bay and out among the islands with visibility that of the interior of a bale of cotton, and felt his way into land-locked Burnt Coat Harbor, where he waited out the storm. He had been given up for lost at Rockland for two days before he finally brought his boat into that port.

But luck has something to do with such a stunt and luck was not with Captain Kent. He missed the channel by a few feet and the boat smashed onto the sunken rocks. Her whistle screamed out distress signals and soon fishing boats were swarming about and taking off passengers and first-class mail.

The whole island was alerted and, as is the island custom in an emergency on the water, almost every able-bodied fisherman was standing by in the bitter evening to give aid. "The night the *Bodwell* went aground" is still a time-mark in the lives of all islanders beyond middle age. At midnight the crew abandoned ship after all efforts to free her had failed. The following day her superstructure was washed away and soon the hull slid into deep water. Though at first pronounced a total loss, she was finally raised and towed to shallow water at Minturn. Later she was towed to Rockland, rebuilt, and put back into service. She had been insured for forty thousand dollars.

But the *Bodwell* was not long for this world. At 8:30 on the night of March 23, 1931, fire was discovered near her boiler as she lay at Swans Island for the night. The crew members still aboard had barely time to escape. The wooden ship went up like a torch, setting the dark water of the harbor and the sombre fringes of the spruce clad shore aglow. Islanders came to watch, and several small boats managed to "get a line on" the flaming vessel and tow her away from the wharf and across the harbor where she sank sizzling to the bottom. She was a total loss except for a few accessories like the pumps. Her whistle was saved and was later installed on her successor, the *North Haven*.

The *North Haven* was the last of the steamboats to serve Swans Island and one of the last in the whole Penobscot region. She was bought in Nova Scotia, where she had been serving Cape Breton. By the summer of 1931 she was shuttling back and forth from Rockland to Swans Island with Captain Kent at the wheel. She was sturdy, with twice the power of any of her predecessors. A pilot house mounted on stilts higher than the rest of her superstructure and far up towards the bow detracted from her looks but added to her efficiency. The house had been constructed in this manner to lift it above the thick lower layers of Nova Scotia mists. It had the same advantage on the fog-blanketed waters of eastern Maine. In rough weather the house in its lofty position swung back and forth in such mighty arcs that it was facetiously estimated that her pilot traveled ten extra miles on the forty-mile run, which in winter took in Vinal Haven.

In 1941 the *North Haven* was commandeered by the United States Government for war service. She served for a time carrying shipyard workers across Portland Harbor. She never returned to her old run. The steamboat phase of the island was finished. A period of transportational chaos followed.

3

An island's boat is so important that she is often the cause of prolonged wrangles. What sort of craft should she be? Big, small, gasoline, diesel? What port should she land at? Rockland, Stonington, Bass Harbor on Mt. Desert Island? Where should she dock on the island? At Minturn, Old Harbor, or Atlantic? What will be her schedule? What rates should she charge? Who will be her captain? The steamboats had become a long-established institution, accepted by several generations of islanders as a matter of course. So long as one was running a regular schedule to Rockland, few questions were asked or complaints launched. The boats were large enough to carry all sorts of freight, including automobiles. Above all they were dependable.

Then suddenly the tie was broken. Rockland, almost a second home for many islanders, became just another city thirty miles away by water and a hundred miles if one went ashore at Bass Harbor and took the land route around Penobscot Bay. On one thing everyone agreed. There had to be another boat, if for nothing else than to carry the mail, which the government would see to in an unsatisfactory manner, perhaps, if the islanders didn't act quickly. But there was angry disagreement as to where the boat should touch on the mainland. Rockland, of course, had advantages admitted by everyone. "When you get to Rockland, you are somewhere. You have a city with stores, bus and train connections, doctors, hospitals, lawyers, schools, everything you need. You dock the boat anywhere else, and where are you? Some little village that doesn't have much more to offer than Swans Island."

The reasoning was sound, but who would finance a boat to Rockland? One of the lobster-buyers of the island, realizing the importance of the Rockland service, made an attempt. For a short time during the war he ran a large gasoline-driven launch, a converted yacht, along the old steamboat route, making stops at North Haven and at Stonington. But in the meanwhile other arrangements had been made for the mail, and without that source of support the Rockland run soon was given up. All are agreed that a boat must have the mail to be successful.

Except among the more inveterate dreamers the Rockland ideal slowly faded. The other islands along the route had got their own transportation and thus removed any chance of much needed cooperation. North Haven had its own boat and Stonington for some time had been connected with the mainland by the Deer Island bridge. Mail, freight, and passengers were already being carried from Stonington to Swans Island on diesel-engined sardine boats. Strife now flared as to whether this was the proper mainland base or whether the boat should run to Bass Harbor on Mt. Desert Island. The reasoning on neither side was compelling. Both places are on islands connected with the mainland by bridges and the nearest railroad to both is at Ellsworth. Bass Harbor is a little closer to Ellsworth than is Stonington, but Stonington is a somewhat larger center than Bass Harbor or nearby Southwest Harbor. The fact that there were no decisive arguments made the battle all the more acrimonious. Stonington, being on the old steamboat route, had an emotional appeal for those who kept their faces resolutely turned westward. But as the issue became hotter, alignments were made on the basis of long-standing rivalries and tensions. Any grudge, however ancient, would swerve a man from one side to another.

In the meanwhile the Stonington service was turning out to be unsatisfactory in many ways. Since the boat was not based on the island, one couldn't leave and return the same day. The connections with Ellsworth were so bad that summer people traveling by train seldom used the Stonington route but went to Bass Harbor by a complicated chain of bus and taxi rides and from there hired a fisherman to take them to the island. Various islanders had built up side businesses in private ferrying of freight and passengers back and forth to Bass Harbor.

An old scow was dug out of the mud at the head of Mackerel Cove and with no repairs put to work carrying cars across the Bay with a motorboat towing it. To say this was a stop-gap would be estimating it too highly. One would ponder long before entrusting a vehicle worth more than twenty-five dollars to this precarious conveyance. The car had to be driven up two sagging planks onto the beached scow and it had to be disembarked in the same way. The trip across, which could be made only in the most halcyon weather, was an hour-long battle for two men with bilge pumps. With luck and back-breaking labor the deck of the scow could be kept an inch or two above water and the car brought safely to land.

Finally a number of "local businessmen," small store-keepers and fishbuyers who upheld the Bass Harbor banner, grouped together to purchase and commission the boat that is now serving the island, the *Sea Wind*. She is Maryland-built, a former yacht, fifty feet long and powered at first by two gasoline motors and later by diesels. From the beginning she has made stops at Frenchboro, a little lobstering village on nearby Long Island, whose inhabitants, never before having had any service, are among her most ardent champions.

But the *Sea Wind* has had a very stiff fight. For several

years she ran in direct competition with the Stonington boat, which still had the mail. Thus for a time the island had two boats, as it had fifty years before. But it got small joy of them. Certain storekeepers staunchly ordered their freight to be delivered by way of Stonington while their competitors naturally switched to the Bass Harbor route. Enemies of the new boat, as they watched her leave port, were heard to heap curses on her, wishing her a speedy trip to the bottom, passengers and all. Nasty rumors, unfounded of course, of attempts at sabotage—a chain stretched across the harbor mouth to wreck her—were circulated. Summer people who used the boat because of the convenience of her connections—she ran twice daily from June through August—were assured by her detractors of her utter unseaworthiness. A man who had just entrusted his own and his family's lives to the *Sea Wind* would be solemnly informed that he was lucky not to have been drowned. The harangue would run something like this: "I wouldn't set foot in that tub. Every plank in her is rotten. She ain't built for these waters."

The *Sea Wind* actually is a most seaworthy craft, fully licensed and inspected by the United States Coast Guard and equipped with every safety device, including two-way radio and fathometer. Despite the malice against her she slowly made her way under the stoical and expert seamanship of her first captain, who suffers from the handicap of coming from "away," and being connected with the island, now his home, only through his marriage. Eventually the *Sea Wind* got the mail contract, the Stonington boat withdrew, the propaganda campaigns were suspended. Judging from the crowds to see her off and greet her, the island has completely accepted her. There is a minor traffic jam at the Minturn wharf at boat time. Still she doesn't take the place of the old steamboat. She can't, for example, carry automobiles. They are transported in a variety of ways—by sardine boat, by the Maine

Sea Coast Mission boat, the *Sunbeam,* or even by the old scow. Also, when you land at Bass Harbor you are in a hamlet of a hundred or so inhabitants. There is still a four-mile taxi ride before you get to the bigger town of Southwest Harbor.

Boat troubles are endemic to the Maine islands simply because boat connections are so vital. Islanders at times develop almost a paranoiac proneness to suspect that the operators of their boats are in conspiracy against them. Yet once a boat and her captain win their confidence their attachment becomes strong and sentimental.

Many fishermen, of course, provide their own transportation in their own boats. Stonington, a waterfront town, is still a popular trading center for everything from groceries to liquor. Many of the islanders keep cars on the mainland so that at any time they can cross the Bay and go their independent ways ashore. Hydroplanes have been used successfully on some islands, but not to the extent of eliminating boats. At Swans Island only one plane has been owned and that for a very short time. Its chief use was to fly ice cream across from the mainland before there was electrical refrigeration on the island. Once it was landed, the owners would dash about in an automobile frantically trying to sell the ice cream—a rarity for many inhabitants—before it melted.

A Bass Harbor fisherman still does a good business taxiing passengers across the Bay in his thirty-foot workboat at the rate of five dollars a trip. Not confined to a schedule he will cross at any time that weather permits. He will land anywhere on the island, and since he doesn't have to follow the round-about route that takes in Frenchboro, he makes the crossing in a half hour—one-third the time of the *Sea Wind.* Yet he in no way supplants the bigger boat. He cannot carry freight or buck the winter gales. He cannot fill the islanders' need for a regular connection with the main. The *Sea Wind* is an emotional, as well as physical, necessity.

8

Summer People and "Natives"

THE STEAMBOAT BROUGHT MORE THAN THE MAIL, FRESH vegetables, and regular communications to the islands. It also brought the summer visitor—"the rusticator," as he is still called in eastern Maine. Bar Harbor on Mt. Desert Island had become a summer resort as far back as the Civil War, thanks to the steamboat connections with the Boston boats and later with the railhead at Rockland. By the end of the century Bar Harbor vied with Newport in fashion and wealth. Thomas Cole, founder of the Hudson River School of painting, first revealed to the nation on his canvasses and by word of mouth the enticements of the climate and landscape of Mt. Desert. Novelists of the period, like Howells in *The Rise of Silas Lapham,* spoke of Maine coast summerings and trips Down East by boat from Boston and New York. Harriet Beecher Stowe and Sarah Orne Jewett in such widely read novels as *A Pearl of Orr's Island* and *The Country of the Pointed Firs* enthusiastically sang the praises not only of the scenery but also of the fisherfolk living in it—their way of life, their speech, and their eccentricities. Miss Jewett said that she considered it the sacred duty of writers to make the various parts of the nation known to one another. She and others like her were highly successful in their efforts.

Mawkish and untrue as many of the literary effusions about the Maine fishermen are, they still have been effective publicity for the vacation business. The overall message of the writers and painters is that a new paradise of lovely scenery and weather and sturdy, old-line Americans of pure English stock exists to the eastward and awaits the delectation of anyone willing to pay to get there and enjoy its pristine blessings.

Some of the islands fared better than others in this beginning of what the Maine Publicity Bureau now refers to as the Recreation or Summer Industry. Mt. Desert, of course, got the lion's share. But Isle au Haut soon fostered an exclusive "colony" that effectually elbowed out the original inhabitants. North Haven attracted wealth in quantities second only to Mt. Desert. But other islands were ignored by fashion and money. Some indeed were ignored by summer people of any sort. Outer Long Island has to this day not a single rusticator. The inhabitants can't decide whether the lack is a blessing or a curse. The feeling at present is that a few summer people might liven things up a bit and bring in a little money.

Swans was long among the islands with only a limited appeal for summer folk. For years the Bar Harbor steamers smoked by without dropping off one vacationer. Finally a few drifted in, but they were not much to speak of. They were college professors impoverished in everything except enthusiasm for nature and "the simple life." But one summer visitor attracts others, usually from the same locality. Pennsylvania, particularly the faculties of one or two colleges in that state, sent many of the early vacationers.

If the islanders were expecting great inpourings of wealth they must have been sadly disappointed. These people didn't spend their days striking pretty attitudes on tennis courts or skimming over the water in sleek yachts. Instead they went to work to earn money to add to their incomes. One family raised berries and apples which they peddled in the villages. Others collected sea urchins and starfish, which they embalmed in formaldehyde and sold to biological supply houses. A theology student started a boys' camp on a basis he hoped would be paying.

That people should have to work for a living seemed very natural to the Swans Islanders but quite puzzling in the light of what their Mt. Desert neighbors must have told them of the ways of "westerners," as they called anyone from beyond the Penobscot. Yet it was perhaps a fortunate introduction for the islanders to have had. These summer people were not the type who, in Miss Jewett's words, would mistake a selectman for a tramp just because he happened to be crossing a field in his shirt sleeves. They did not by a show of idleness antagonize the natives who themselves had to work, whether it was summer or not.

A daughter of a Maine coast farmer, now a teacher in New York state, says that fifty years ago when she was a girl the term "summer boarder" was the most insulting thing one child in her community could call another. The annual arrival of the vacationers used to be called the "summer complaint." The present semantically neutralized term, "summer industry," implies a more moderate attitude. The hordes of tourists and cottagers are now regarded as a means of livelihood, a seasonal phenomenon like the schooling of herring and the spring crawl of lobsters in from deep water. Yet even today many a local resident in Maine—a person of lineage as ancient and distinguished as any in America—harbors a bitter dislike of summer people planted there by some vulgarian years ago. One Maine island girl, now a noted novelist, tells of being confronted as a child by two tourists who, eyeing her with astonishment, exclaimed: "That must be a native." At once the child dropped down on her hands and knees and began to roar like a beast. She was going to give the summer folk a real show. And indeed there is something offensively condescending in the terminology used by summer people: "native" and "summer colony"—an outpost among the barbarians. Yet today these terms have lost most of their sting. They are used by summer people and natives alike.

For a time on Swans Island a number of boarding houses—euphemistically called hotels—entertained two hundred guests a year. This was in the rocking-chair era, when an American's idea of vacationing was the restful one of sitting on a veranda and absorbing the view and local color. The one island "hotel" still in its original form—though now a private home—has a broad porch on three sides with extensive views up the Bay. Here the boarders used to sit and admire the scenery and exchange pleasantries with the sea captains (any man who owned a boat larger than a dory) as they drifted by on the board sidewalks that were then the pride of the island. Built in the nineties, the establishment was for some reason called The Ponceana, a name under which it still labors.

Another hotel, at the harbor, catered to vacationing officials of the Pennsylvania Railroad. The story is told that these gentlemen so valued the seclusion of their chosen hostelry that when other guests appeared they came to dinner dressed like ragpickers to scare the newcomers off. This hotel has long since gone out of business and in fact, in the words of the islanders, it never operated except "by guess and by damn." First its upper floors were removed and the remaining ground floor served as a pieshop and restaurant and, on occasion, as a dance hall. Now even the lower floor is collapsing. The remains of one two-story wing are occupied by the head selectman and his housekeeper.

The most enterprising bid for the summer trade was Idylwild, about which the islanders love to talk. It was located at Irish Point six miles from the steamboat wharf by way of almost impassable roads. "There wasn't much choice between the rocky roads to Dublin and to Irish Point," a local wit said. But in order not to be dependent on such hazardous land travel, the proprietor bought a tiny steamboat, "not much bigger than a punt," behind which he towed a seine dory loaded with his boarders. When they were not rocking on the veranda he would convey them fishing or sightseeing among the islands, or to Old Harbor where, as one eyewitness describes, "they would land and run about to their hearts' content." But Idylwild also soon closed its doors. Later it was torn down and its timbers are now built into a dwelling.

2

The fling at the rocking-chair trade was short-lived. One or two women still take in boarders and lodgers but only in minute numbers. Today the summer people live almost exclusively in their own cottages. Of the more than two hundred dwellings on the island they own perhaps thirty. Some of these are scarcely ever occupied. Of the vacationers who come regularly, several remain as long as six months; others for only a week or so. The usual stay is about one month. Furthermore, with but three or four exceptions the summer persons are still of the most modest means, and the few who do have money choose to live simply. Many are old maids—so many in fact that the continuance of the colony may well be considered in jeopardy. There are only four summer families with children. As it has been from the beginning, many of the cottagers are teachers. At one time two of Columbia's most distinguished professors had houses on the island. Three or four other colleges and the public and private schools of several states are represented now.

On Swans Island all resident property is assessed at \$154,150; non-resident at \$55,920. Some of the non-resident is owned by natives now living away or by lumber companies and the like. The property of summer people does not account for more than \$35,000 of the non-resident valuation, or one-sixth of the total for the town. Compare these figures with those of the more fashionable Isle au Haut. The total valuation for that island is $132,522, of which $104,436 is non-resident. Here again the summer people do not own all the non-resident property, but they own enough to make their holdings about two-thirds of the entire valuation. On Isle au Haut more than half the houses are empty in winter. Though the fifty or so all-year-round residents do some fishing, the summer people are a major source of income. They return year by year with greater certainty than either lobsters or herring, and they are not so subject to vagaries of the market. They give the island an economic stability that a community dependent on fish does not have. The difference shows up graphically in the tax collector's reports. In the year 1944, for example, when the collector for Isle au Haut listed $5.97 extant in unpaid taxes for that and the preceding ten years, the corresponding figure for Swans Island was $2,047.26. Summer people not only are better able to pay their taxes regularly, but a town that has many summer people has a resource that will help keep its residents solvent through the ups and downs of the fish market.

Thus at present on Swans Island the summer people are neither a major problem nor a major asset. They annually pay about thirty-five hundred dollars into the town treasury. They give a little trade to the stores and to the boatmen who ferry them across the bay. They supply work for one or two painters and carpenters, if those artisans feel like working. But most of the year the summer people are conspicuous by the absence of their persons and their influence.

Though the storekeepers lament that the vacationers are on hand in such small numbers and so briefly, the island as a whole, if it wishes to remain a fishing community, is fortunate that the trickle of outsiders has never become a flood. Floods of that sort sometimes succeed in inundating the whole local way of life. Even Bar Harbor was once a quiet fishing village. Some would argue that what it is now is better. They would say it's a wealthier, livelier community. But others would argue that too many of the old-line people have the status of servants of the rich pleasure-seekers who have taken over the town. George Wasson in his *Sailing Days on the Penobscot* speaks of the degenerative effects the summer colony on Isle au Haut had on the original residents. Robust young fishermen belied their heritage by going ashore to be chauffeurs for summering millionaires rather than following their manly calling of the sea. Of course, Wasson may be a sentimentalist, and driving a Lincoln for steady wages may be preferable to plunging about in a smelly boat trapping uncertain lobsters for very uncertain pay. Many Maine youths have taken this view; yet many others, and perhaps Americans in general, still labor under the prejudice that working for oneself is a little more satisfying than working as someone else's menial.

Certainly Ruth Moore, a native of the islands, deplores in her novel *The Fire Balloon* the effects of what one of her characters calls "sucking up to the summer people." As a matter of fact Miss Moore, whose girlhood was spent close to the millionaire colony on Mt. Desert Island, regards with indignation and contempt the snobbery, callousness, and stupidity of many of her summer neighbors. Mary Ellen Chase, who witnessed the transformation of her native Blue Hill into a summer resort, writes of the same problem. In *Mary Peters* she regrets the rift "between sojourners and natives" which grew until "two alien peoples share the coast." She saw "something immeasurably sad in the sight of a grandson of a shipmaster in the foreign trade shingling the roof of a summer cottage for his livelihood." In part she held the summer people at fault because "they knew little and cared less for the boy's history." The indifference of these newcomers to Blue Hill was complete. "They didn't discriminate socially against the natives," Miss Chase points out. "They simply ignored them. Yet the villagers were perhaps ultimately responsible. They were all too ready to fall into a servant-master relationship with the rusticators." She sees a "shrewdness in the New England rural mind . . . steadily developed through two centuries and more by struggle with land and sea, by the necessity for thrift and quick thinking and the seizing of opportunity. . . . Under the best of circumstances, under the mellowing influences of contact with other lands and other minds, that shrewdness lost itself in a humorous kind of wisdom. . . . Under more narrowing circumstances it doubled back upon itself, became more acute, subtle, cunning even." In other words, the exploiters were the natives as much as the summer people. And relations between exploiters and the exploited are likely not to be cordial.

The barrier between the summer and all-year-round people at Blue Hill and Mt. Desert Island was and probably is much more formidable than on Swans Island, where the "sojourners" are less pretentious and less numerous. Yet the division is very real here, too. The ultimate test of mutual acceptance is intermarriage, which among summer people and natives of the Maine coast—and of the rest of New England as well—is almost unheard of. No pairing of a summer person and native has ever occurred on Swans Island. If it did occur, it would cause as much dismay and tongue-wagging as an interracial marriage in a deep Georgia hamlet. The results would not be the same, of course. There would be no threatened lynchings or ostracism. Love affairs have developed between natives and rusticators, but have never prospered. It would be wrong to say that public opinion is too strong to make such a marriage feasible. Something much less articulate than opinion comes into play; some deep-seated group feeling asserts itself through the individuals themselves.

Rural New England boys and girls when they go away to

cities to college or to work are altogether at ease with the same people whom they would distrust and be stand-offish with if they knew them as summer vacationers back in the mountains or on the coast. For almost all these summer people had their origins, a few generations ago, in the type of environment they now seek their recreation in. The summer person and native are usually of the same stock—overwhelmingly Anglo-Saxon—with the same political, social, and religious ideals and mores. Students of discrimination and group tension should take notice of this fact. Race, creed, even economic status are not at the bottom of this very real, sometimes very bitter, antagonism. Negroes enjoy much more social intimacy in these New England villages than do their white employers. A Negro servant will be referred to very sincerely as a "colored gentleman" and be invited to spend his day off in the family circle of a native whose dinner table has never known the presence of a rusticator. On Swans Island the attitude towards a Negro is simple, "He's a human being, isn't he?" But no native has ever been heard to say of a rusticator, "He's a human being, isn't he?" Whatever the islander's views on the matter are, he keeps them to himself.

At any rate, at no time on Swans Island have numbers of the population been in danger of being reduced to menials. The really cushy jobs—sailing yachts and chauffering—are not available. Not a single pleasure craft larger than a catboat is owned on the island. Most of the summer people have no cars, and if they do, it is some antique like a 1935 Dodge or '39 Ford—far more disreputable than the vehicles the natives own. A few islanders have taken jobs as house servants, but not as a regular thing over long periods of time. The summer people at Swans Island, whose incomes are less than those of many of the lobstermen, usually can afford to hire no one, even if willing help could be found.

3

There are naturally curmudgeons among the islanders, and meddlers and troublemakers among the cottagers. Those summer people who attempt to run island affairs even in the smallest degree are met with insuperable opposition. As one island woman put it, "We don't mind the summer people so long as they don't tell us what to do." The attitude is a bit surly. Relatively few summer people have ever attempted to tell the islanders what to do in important matters and the advice of those who have has never been heeded. One man, the begonia grower, who made his living on the island for twenty-five years but was still somewhat perversely regarded as a vacationer, wished to express his gratitude to the community by presenting to it a fully modernized residence with an operating room where the island physician could live and practise. The house had plumbing, a home electric plant, and central heating. The island had always had trouble keeping a doctor because of lack of just such facilities. The new house was to be given to the town with the stipulation that it be used as the doctor's dwelling. But because of this stipulation, the town fathers and the voters turned it down. They wanted nothing with strings attached. Of course, they paid a price for their independence; the town was for ten years without a resident practitioner.

Instinctively the islanders know that if they let too many things be done for them they will slowly but certainly lose their independence. In their work they have been entirely on their own. Perhaps their business, fishing, is the only industry that has less government interference today than ever in its history. Price supports, though voted by Congress, have never been applied. But summer people have threatened another kind of interference. Among them the desire "to better the natives" has been rife, as is to be expected in a group of elderly school marms whose entire lives have been spent molding children into civilized beings. Their island efforts range from attempts to institute handicrafts and libraries to campaigns for the betterment of private morals. None of these ventures have met with success. The islanders hire teachers, nurses, doctors, ministers for these purposes.

Several flare-ups have occurred when a summer person has attempted to break some established custom, such as the disposal of garbage on a certain beach which he has bought and now wishes to keep free of tin cans, banana peels, floating cardboard boxes, and orange rinds. One islander complained that one Sunday he was unable to find a stretch of the shore where the owners would permit him to picnic. "I felt like a stranger," he said. "Some summer person had posted every place I went to. I felt I didn't belong here." He didn't need to add that his family had been on the island for one hundred and fifty years. But such complaints are rare. On Swans Island more trouble over rights-of-way and shore privileges occur among the natives than between summer people and natives.

The summer people are in their ways as individualistic as the islanders. Had they craved a gay social life, they would not have come to Swans Island. Nor would they have come there for an easy vacation. As one of them said, "Summering on Swans Island is not the path of least resistance." One must be able to help himself, for no one else will have the time or inclination to help him even for pay. There is a feeble pretense of a social life among the summer people, but no large-scale entertaining either at cards or cocktails.

Most of the summer people have their own preoccupations. A few—only two or three—have motor or small sail boats. Some play archeologist a bit, scratching about in the numerous Indian shell-heaps. Some indulge their whim to capture on canvas the surrounding natural beauty. Others are botanizers and bird watchers, who scramble miles through underbrush to find a twinflower or a cormorant's nest. The greatest number make a full-time occupation of the beautification of their homes and ground. All are professional islanders, delving into the island's history, conversing endlessly on its present problems from the inconvenience of the mailboat schedule to the increase in shotgun marriages. Those who have cultivated some friends among the natives become vainglorious, assuming themselves to be real islanders in contrast with those who can boast no such acquaintances. Rivalry exists as to who is on friendly terms with the most natives.

More than one islander has commented, however, that the really good times on the island take place after the summer people leave—a remark frequently heard around summer resorts. The implication is that the summer people inhibit the fun-loving instincts of the natives. This, however, is not the case. Summer people are always welcome at such affairs as church suppers and sales. But summer people and natives seldom get together for an evening of cards or gossip. Sometimes they join forces for a bout of drinking, but these contacts are between non-conformers in both groups.

In general the summer people enjoy the physical and human attractions of island life and take the drawbacks in their stride. But two small groups are much more violent and vocal in their reactions. One of these sentimentally pronounces all island people and all island customs, all meteorological and geographical features of the island to be altogether lovely, superior to everything else in the universe, a paradise regained. They come to the island to escape. Some are persons who have had strong family attachments now broken by death. The island has become associated with a brother, a son, but more often a mother, with whom years ago they spent their summers here. The retreat is not only away from the world but into a safer, happier period of their lives, youth or childhood when someone shared their cares or assumed the full burden of them. These people can have their windows smashed by local rowdies, their wells defecated into, and promised jobs postponed for years, but they will still see nothing but sweetness and light in the collective island character.

The other group, while agreeing on the scenic attractions of the island, cannot find enough unfavorable comments to make on the tone of island morale. There is a typical state of mind for such a person to be in. You will find him of a sunny June morning storming about his yard cataloguing in a voice sufficiently raised to be heard by any nearby islander all the vices of this shiftless population among whom he has to spend his summer. One can't get a carburetor adjusted; there is no mechanic on the island. The one who used to be here left because he couldn't stand the place—a point of view with which this summer person strongly sympathizes. One can't trust an islander anyway. He will promise to clean out one's privy but will not show up for months. The stores are no good—prices exorbitant, food stale, selection atrocious. The islanders are laggards, ne'er-do-wells, degenerates. Where is their civic pride? How can they expect to have summer people if they accord them such shabby treatment? No wonder all the up-and-coming islanders left long ago. The tirade continues day after day, week after week, till at last the summer has vanished—the carburetor unadjusted and the "little house" overflowing.

Which is the more devoutly attached to the island—the adorers or the denouncers—would be difficult to say. In both groups are rusticators who have not missed a summer in forty years.

4

No one has adequately studied the effects on the nation as a whole of the mingling of rural and urban persons in the resort areas of the United States. Still, such a widespread American custom as the summer vacation must have marked cultural results. Consider the case of this one Maine Island. Summer after summer teachers from a half dozen states come to it, live on it, mix with its people. In the fall they return to their jobs, where they come in contact with hundreds and thousands of the young of the country.

What do they bring back to their classrooms, to their communities, and to their own daily living from these island summers other than rested minds and bodies? They bring a little of the repose of the spruce-covered hills and shores, the gladness of the bay at noon when the gulls dip and circle above the white-crested waves, the splendor of the water aflame at evening as the sun drops behind the western headlands, and the world-obliterating solace of the fog in late afternoon rolling in over the islands. A summer person fresh from one of the metropolises will put the matter thus: "Somehow when a fog settles down on you here you feel removed from the mess the world is in. You can forget atomic bombs. You feel secure." It is a blunder to sympathize with a vacationer whose arrival is greeted by a spell of fog. "Don't sympathize," the answer may be. "I like the fog better than anything else. That's why I come here."

The island fills an emotional need for the vacationers. They store feelings that stay with them and support them in the frustrating days that must pass before the next island sojourn—

> feelings too
> Of unremembered pleasure: such, perhaps,
> As have no slight or trivial influence
> On that best portion of a good man's life,
> His little, nameless, unremembered acts
> Of kindness and of love.

Perhaps these are what any vacationer seeks who flees to nature. Yet the island visitors come to live not only in nature but in a human community. They think this a simpler and kindlier community than most of those they have lived in on the mainland. Some of the older decencies (one would like to quote "virtues," but he does not wish to be derided as a sentimentalist) the vacationer believes linger here—a respect for one's neighbor, an innate courtesy, an independence of spirit which at times may be irritating but is a basic asset, and above all a simple, open warmth elemental in all human beings but too often chilled in more sophisticated societies.

The natural scene is more meaningful because of these people who live in it. With their reassuring villages in the coves and at the foot of the hills they have softened the harsh, savage beauty of the granite shore. Their sharp-roofed houses in the fields and clearings cheer the bleakness of interminable forests of spruce. Their white boats and many-colored buoys lighten the indifferent blues and grays of the water. Many places are beautiful because man has been excluded from them. This coast owes much of its beauty to the fact that human beings live here.

The spectacle of a vigorous, independent, versatile, and intelligent group of people successfully and decently carving out a way of life for themselves in continuance of an ancient and distinctive tradition is calculated to have the effect of bolstering one's confidence in humanity and the nation. If some of that renewed assurance of the beauty of the world and of human life is carried back into some mirky classroom in Brooklyn or lecture hall filled with youthful cynics from our "better" families, then this island community and many others like it are contributing in a subtle but essential way to the happiness and understanding of the American people.

But the benefits are not always one-sided. The summer people need not be parasites drawing spiritual sustenance from the island and leaving nothing in return. During the last century most Maine coastal towns escaped ingrowth because many of the inhabitants were constantly sailing up and down the seaboard and about the world in their fishing and merchant ships. Their horizons had little chance to become restricted. But with the decline of shipping and fishing, the remoter communities ran the danger of becoming almost entirely shut off and degenerating like some settlement deep in the mountain coves of Kentucky or Tennessee. Some of this sort of degeneracy has developed in Maine, on the islands as well as in the interior, where the tourist can see it in the shanties strung along the through highways. But it could have become much more serious had not the world started coming to Maine at the very moment that Maine stopped going to the world.

5

The future of the "summer industry" on Swans Island looks bleak. Transportation to and from the island is difficult and rather expensive. There are no attractions for the usual motoring American tourist—no historic monuments or "amusements." The type of person who likes the seclusion that Swans Island has to offer can't afford to maintain a house there except at great sacrifice. One phase of the summer industry, camps for boys and girls, has never flourished, partly because of the island's inaccessibility and partly because the great drawing card and time-killer of such camps, swimming, must be highly restricted in the Labrador Current that keeps the water in the fifties even on the warmest days of summer.

Had the regular steamboat service with Rockland been retained, Swans Island might have a chance of holding its own as a summer resort. One could drive to Rockland, put his automobile on the boat, and after a morning on the water be at his summer cottage. With the loss of the steamboat, Swans Island lost its chance for a sizable cut in the summer trade.

However, there will always be cruising yachts. Each summer ketches, schooners, and yawls swarm down through the islands on their voyages eastward. Like motorists on the main highways, they seldom linger anywhere. They pass by the harbors, villages, and towns, but remain strangers to the slightly aghast dwellers along the route. Their destinations are always the same—the famous resorts like Camden, Northeast Harbor, Bar Harbor—or, for the bigger craft, the more remote and alluring coasts to the northeast: the Bay of Fundy, Cape Breton, the Gaspé. When a lobsterman passes near one of these boats he returns the gesture or spoken word of greeting with the camaraderie of the water. The pleasure boat scuds on, dipping gracefully into the waves, masts slanting before the wind. And the lobster boat, dirty and noisy, wallows on to the next buoy.

When a yacht puts in at a village for provisions, its call is brief. When the purchases are made, she slips silently out of the harbor, or if it is evening and time to lie to, she skulks over against some uninhabited stretch of the harbor shore, near the entrance if possible, or seeks out some deserted cove to spend the night in. So the yachtsmen and islanders remain aloof, but they are not enemies. There is mutual respect, for the yachtsmen are expert navigators, almost all of them. They are courteous and considerate on the water. They are careful not to foul and cut the lobster buoy lines with their deep keels, and they always leave plenty of room for a working boat.

Sometimes cruising schooners from Sedgwick, Maine, anchor for the night. These are old coastwise boats made over to accommodate several dozen vacationers who wish to pass a week or so of sailing under old-time conditions and are willing to help with the operation of the vessels. In the evenings when they drop anchor, the passengers come ashore and spend money at the local stores or patronize any entertainment that may be scheduled.

Occasionally the aloofness of the yachtsmen dissolves and a memorable time occurs. A number of craft may be anchored for the night in one of the harbors. If there is a lobster supper or dance going on ashore, the yachtsmen will be attracted to it. They spend money, eat voraciously, and mingle with the islanders. Or perhaps the people from the little fleet of cruising boats that anchors in the island harbors will bring ashore their musical instruments and an impromptu dance will get under way in a store. Before long the floor will be jammed with vacationers and islanders mingling unselfconsciously. The party will go on till dawn, and be talked about many years after.

9

Requiem of the Winds

IN LITERATURE ISLANDS HAVE ALWAYS REPRESENTED HAVENS from the harshness of the actual world. The mythical Atlantis, Homer's Phaeacia, where Ulysses found relief from an almost intolerable life, More's Utopia, and Melville's Typee have all been happy because insulated from the world at large. Other islands like Avalon have been the abodes of death, the surest escape of all from the struggle of existence. The imagination in harmony with man's deepest needs yearns for security, for protection from hate, fear, and despair.

In Elizabeth Ogilvie's books of a Maine island—*High Tide at Noon, Storm Tide, Ebb Tide*—the Bennett's Island cemetery always looms mistily in the background. It is located in the center of the island on the far side of an apple orchard long ago gone wild. There beneath the perpetual twilight of virgin spruces, whose tops sway ever so gently in the ceaseless southwest breeze, are the graves of the island dead and of two or three unidentified sailors washed ashore after a storm. It is the most peaceful place on the island, so sheltered from the sea that the rote of the waves is dulled to little more than a lullaby. The graveyard is a place of surpassing peace, where the hush of solitude suggests the most shadowy recesses of a cathedral.

Carved on a memorial stone at Frenchboro, the plantation neighbor of Swans Island, the following epitaph commemorates a young captain and a boy who were, as the inscription further relates, "lost on board the schooner *Henry Clay* of Tremont in the great gale of September 4, 1851, at Prince Edward Island."

> No village bell shall toll for them
> Its mournful solemn dirge;
> The winds shall chant a requiem
> For them beneath the surge.

A Swans Island seaman, Joshua Trask, was also lost on the *Henry Clay* in the same disaster, which took the lives of the entire crew. But news of the sinking must have been slow in trickling southward, and corroboration must have been slower still—and desperately painful. The inscription on Joshua Trask's stone—"lost in the Bay of Chaleur in the gale of October 3, 1851,"—is quite different from that at Frenchboro. Perhaps there had been two storms that autumn in the Gulf of St. Lawrence, so that no one knew in which gale or in what part of the gulf the *Henry Clay* had foundered. A month was not such a long time a century ago. Since no one survived the wreck, there was probably no eye-witness to it. Hearsay, conjecture based on bits of wreckage picked up by other fishing vessels, were all there was to go on.

Thus two adjacent islands have two widely divergent versions of the calamity. But it mattered little then, and less now. What did matter was that Joshua Trask, aged forty-four, left a widow and six children on Swans Island. The snatch of doggerel still legible on his weathered tombstone hardly does justice to the facts:

> Farewell my wife and children all,
> From you a father Christ doth call;
> Mourn not for me, it is in vain
> To call me to your sight again.

These commonplace lines take on some of the cleanness and sincerity of true poetry when read in a hill-top cemetery where the unkempt grass bends to the salt breeze and the backdrop is a panorama of islands and ocean stretching out to a glittering horizon.

The same quatrain appears on an overturned stone in a family plot deep in the interior of the island. The picket fence—a pathetic attempt to render inviolate this spot of death—is tumbled down too. Nearby is a cellar hole full of the rotting timbers of the family dwelling. In the surrounding field are gnarled and decaying apple trees, their fruit untasted for generations except by deer whose tracks and droppings are everywhere plentiful. At the end of a vista between two hills the white spire of the Baptist Church rises above the black spires of the spruces. In this church the family buried here once sang hymns to God in the same meter as that of the epitaph. This was the poetry these people knew—that of the hymn-book and folk-balladry, and finally of the gravestones.

During the last century and even into the present one, life on the islands was hard and precarious, especially for the young. The preponderance of graves are of persons under thirty, and the rare man or woman who lived to ninety or a hundred must have lost all count of the family funerals. Death had a habit of striking suddenly and mercilessly and in many guises. Until fifty years ago tuberculosis was a scourge referred to as the "island disease." The great influenza epidemic of 1918-1919 took over twenty people in the prime of life. The sea, too, preyed especially on the young and vigorous. In the history of Swans Island well over one hundred persons have lost their lives at sea from Labrador to the Caribbean, and doubtlessly every ocean in the world has received its contribution of victims from this island. Many have died on ships at sea or at some distant port from causes other than drowning. The same stone that records the loss of Joshua Trask bears a notice of the death of his son Napoleon Bonaparte Trask—Joshua was a man of "liberal education with an enthusiasm for French history"—who succumbed, apparently of natural causes, in the port of Sidney, Nova Scotia. Another islander died of yellow fever in the West Indies and another of some sickness in Rio de Janeiro.

2

A number of mass drownings are commemorated in the cemeteries of Swans Island, either by stones to individuals or by cenotaphs for a number of persons from the same family lost in the same disaster. In November of 1860, the schooner *Constitution* capsized in a squall off Nausett light on Cape Cod. The vessel, which was carrying a cargo of smoked and salted fish to Baltimore, was manned by a crew of four, all from Swans Island. The captain lashed himself to the rigging in the hope he'd stay afloat with the wreck, but it soon went down, dragging him with it. One managed to stay afloat on a plank until rescued but the other three perished.

The worst shipwreck loss was that in January, 1867, of the schooner *S. J. Collins*. The toll of lives is chiseled into three sides of a monument in the graveyard in Atlantic. Ten persons, all of them young, were lost: the captain, his wife, their baby daughter, and the wife's daughter by a former husband, himself drowned in the wreck of the *Constitution*; the wife's two brothers, their wives, and a daughter of one of the couples; and an unrelated man, the exception on this family outing. In December the *S. J. Collins* took on a cargo at Bangor and headed south to New York. After a successful fishing season the young people aboard were bent on combining business with pleasure, as was the custom in those days of more spacious living on the Maine coast. On January 5, they started homeward with a load of corn for Belfast and Bucksport. The first night out a gale pounced down and engulfed the ship. No bodies were recovered. The dead are remembered today by an urn-crested cenotaph erected by an aged man who had already lost his first wife and now in a single stormy night had lost three children and their spouses and three grandchildren.

Death by drowning haunts the minds of all islanders, especially the women, who are painfully conversant with the tally the ocean has chiseled into the island headstones. A woman often says she is glad she can't see the water from her house. The view would give her no pleasure, knowing her husband was out in his boat. For other women the trip on and off the island is so great a terror that they prefer not to leave at all. But the island men have learned to joke about the possibility of drowning. After a narrow escape a fisherman will say: "A man who is being saved to be hanged won't drown."

The great marine disasters of the coast are always lurking in the islanders' consciousness. Never a gale of wind comes up but the older people think of the sinking of the *Portland* near Provincetown with the loss of one hundred and seventy six lives. The "*Portland* breeze" of November, 1898, has become a standard by which all subsequent northeast storms have been rated. An oldtimer after a blow from that quarter will demonstrate in detail just how it differed in its effects in such and such a cove or on such and such a headland. The waves were just a little higher on this ledge; the wind was just a shade more in the nor'ard and thus blew three sloops ashore; the tide pushed a little farther up into the woods.

3

On Swans Island are several family graveyards like those described in Frost's poem "Home Burial," but they have long since been overgrown in disuse. On Outer Long Island, though, the custom still holds. A mound of raw earth six feet long is not a surprising sight in someone's backyard. There is a larger cemetery on the island, but many of the houses have a little cluster of stones nearby where the family dead are laid to rest. Swans Island, following a practice widespread in New England towns, voted in 1923 the sum of one hundred and fifty dollars for the upkeep of the three main cemeteries—one at Atlantic, one near Old Harbor alongside the Methodist Church, and one in the "North," the now almost uninhabited area beyond the Carrying Place. The maintenance consists of keeping fences in repair and mowing grass once or twice a year.

Death on Swans Island, as in many New England communities, has become a town affair. The voters annually elect two sextons to supervise the burial of the dead. As far back as 1903, the town meeting voted four hundred dollars to purchase a public hearse and to build a shelter for it. It was a glass-paneled, filigreed vehicle just large enough to slip a box into. The following year money was voted to lengthen the "hearse house." For many years the horse-drawn hearse conveyed the island dead from their homes to whatever cemetery traditionally received the bones of their families. After horses went out of fashion, the old hearse was discarded. Its glassless, paintless skeleton lies alongshore, its wheels long since removed for use on a farm wagon. For a time, in the open seasons, the dead had their last ride in a truck. In winter a work horse sometimes hauled the coffin on a lumber sled through the thickly crusted snow that clogged the roads. When the smaller neighboring islands were inhabited, a floating funeral, a procession of dories, sometimes crept from the landing place in some spruce-fenced cove and slowly filed across the water to the larger island.

In the last few years the town has again owned a hearse. The vehicle was purchased second-hand for four hundred and fifty dollars and a garage was erected for three hundred and fifty dollars. The price of death, like that of lobsters, has risen in Swans Island since the horse-drawn hearse was bought in 1903. There was much opposition to this recent extravagance. One of those who most vociferously complained was the first to ride in it as a corpse, much to the stimulus of the islanders' sense of poetic justice. A licensed undertaker, who is one of the sextons, embalms and buries the island dead according to the requirements of the law. Graves are dug by any able-bodied men available. In one of the cemeteries the ground is so wet that it is hardly suitable for burial. Sometimes the graves there fill so rapidly with water that stones have to be placed on the coffins to keep them down.

Like many small communities, the island has a way of reclaiming its own in death. Persons who have lived away during most of their lives are brought back to the island from distant places. More and more summer people are buying plots in the local graveyards. The thought of the crowded impersonality of a city cemetery is too much for them. In their wills they request that their bodies be shipped from great distances to rest in the fog-shrouded peacefulness of the island.

Sometimes the sea casts up a stranger to be buried in the island earth. The town report of 1905 records that a citizen was paid two dollars "for finding a dead body and reporting same." After a severe storm one hesitates to probe deeply into the debris along the beaches. Years ago a schooner laden with laths from Nova Scotia went ashore on nearby Marshall's Island in a winter gale. The next day fishermen from Swans Island crossed over to examine the wreck. In the ice-caked rockweed on the beach of what was thereafter known as one of the countless Dead Men's Coves along the coast, they found the frozen bodies of the three crew members of the stricken vessel. One man had a fragment of plank driven through his body from side to side. The islanders loaded the corpses into their boat and carried them over to the island. Before they buried them one man cut open one of the bodies to see if a rib was missing. Did religious zeal or scepticism prompt this macabre testing of the story of Adam and Eve? Word of the wreck was sent down to Nova Scotia and soon relatives of the drowned captain arrived. The bodies were dug up and the captain's was identified and taken away. The other two, unknown even to the Nova Scotians, were laid back in their unmarked graves.

10

Religion and Medicine

THE FIRST SETTLERS ON THE ISLANDS WERE CLOSE ENOUGH TO Puritanism to think it the gravest of sins if two or three failed to gather in worship on Sunday and if possible on Wednesday or Thursday evening for a "lecture" and prayers. But Swan failed them here as in everything else. In violation of his agreement with the State of Massachusetts he neglected to build a church, and with no church it was long before a pastor resided on the island. On the mainland pioneer settlers had the chance to go overland to the nearest town to meeting. But in the days of sailboats, family trips off the island for church or any other purpose were major undertakings infrequently attempted. Even with motorboats the difficulties are great enough to make such excursions rather uncommon. So for a generation, midweek prayer meetings and Sunday services with a sermon read by one of the worshippers were held in a private home or during the summer perhaps out-of-doors under a tree. The need to worship was compelling among the people; they came from all over the island by ox team, by water, and afoot. On fair Sundays sails would put in from the settlements and solitary homes on neighboring islands to swell these improvised congregations.

In 1814, the first minister, a Baptist, built a house and settled on the island. He found fertile ground for the religious harvest he hoped to reap. In a series of revivals he made several dozen converts. A year or two later a Baptist Society was formed and taken into the regional conference. Membership flourished. In 1824, thirteen were baptized in the icy waters of the ocean. In 1842 another revival brought in twenty-six members. But within a few years the ardor had burnt out and for a time meetings were discontinued. Then in 1857, a third revival brought in twenty-four new members and for the next thirty years the congregation grew till it numbered more than seventy. In the meanwhile a church was built in Atlantic where its steeple—now removed and replaced by a cupola—became a landmark for boatmen up and down Blue Hill Bay and on the ocean to the east. A succession of ministers came and went, the most distinguished being the President of Colby College, who preached on the island for three summers during the '80's. A roomy parsonage was built and was occupied off and on for thirty years. It is now a meeting house for the Ladies' Aid and contains the minister's office.

By the turn of the century the Baptist Church was already in a decline that nothing and no one has so far been able to reverse. At present attendance averages about twelve—more in summer, when the rusticators are on hand. It has long been the custom for the summer people to sit on the north side of the church and the year-round residents on the south, leaving a section of empty seats in the middle. How the custom started no one knows, but one doesn't lightly break a tradition on Swans Island. The minister gets ample neck exercise trying to address both groups at once. Before and after the service the two groups mingle freely and cordially.

Old Harbor has a Methodist Church whose history is similar—a steady growth after a late beginning, a peak marked by the building of a meeting house, and a slow, unreversible decline to an average attendance of ten or a dozen. In Minturn the Advent Church, the last of the three major congregations on the island to organize, has fared no better.

Ministers do not stay long on Swans Island. One of them said, "We get sick of preaching to empty pews." All of them admit that the churches play a pitiably small part in island life. A few families in each village attend regularly. Others send their children to the Sunday Schools, which in fact are thriving, but stay away themselves. The majority of adult islanders are unimpressed by organized religion as they have seen it. This is true on many of the other islands but not on all. Beals Island, off Jonesport, is a church-going community. Its five churches—two Baptist, two Adventist, and one Latter Day Saints—are well filled each Sunday from a population of only six hundred. One minister estimated that two hundred children attended the Sunday Schools. On the other hand, Matinicus Island for forty years had no regular pastor, though in summer it frequently had a clergyman who, through arrangements with the Maine Sea Coast Mission, got a summer vacation in return for the not very arduous duties of an island parish. Church membership there dwindled to seven until the Mission sponsored a resident minister for a full year. For the first Sunday or so after his arrival, fifty or sixty people came to church out of curiosity. Fourteen eventually joined as regular members.

Part of the decline in religious interest is due to the times, of course. Protestantism has lost much of its hold on the imagination and emotions. For the islander, certainly, it has become too tame. In the last century the Baptists, Methodists, and Adventists made their mark on the islands through revivals, direct frontal attacks on the religious emotions. Mary Ellen Chase describes these awakenings as she knew them in her native Blue Hill. They were yearly events, stirring deeply the spiritual life of the town. Islanders today who never darken the door of a church listen to some of the more energetic Sunday evangelists on the radio. They frankly say they like "old-fashioned preaching." The island ministers themselves realize that if their sermons are "think pieces" they will probably only bore their parishioners. Yet the pulpit-pounding of the old-time tent meeting is distasteful and impossible to clergymen of the type available to the island.

The islanders show a similar taste in hymns. They love rollicking ones like "Life Is Like a Mountain Railroad," "Throw Out the Lifeline," "Let the Lower Lights Be Burning," "Come to the Church in the Wildwood," and "Standing on the Promises of God." Many a dragging service has suddenly been sparked into life by the singing of one of these. The people readily admit they want exciting rhythms in their church-singing, and the minister who selects accordingly is gratified by the response. Often when their craving for quick tempo is satisfied, they will sing all the more feelingly such quieter hymns as "Rock of Ages" and "In the Cross of Christ I Glory."

A religion that offers quick and ample outlet for the feelings has always prospered on the Maine Coast. Adventism at the time it came to Swans Island was charged with emotional and imaginative power. By becoming just another Protestant religion it lost much of its force. Mormonism, too, got an early foothold on some of the islands and set the people aflame with zeal. Missionaries "labored with great success" on Vinalhaven and North Haven—then known as the Fox Islands—and organized congregations there and on the mainland. A group of Latter Day Saints from Beals Island set out for the Holy Land, there to settle permanently in conformance with what they thought was the will of God. Within a year the colonists were back in Maine again, having suffered great hardships. To this day there are strong congregations of Latter Day Saints on Deer Isle, Beals Island, and other spots on the coast of eastern Maine.

The islands have often been targets for missionaries, who intuitively sense the religious potentiality of these lonely spots.

Year after year groups of Jehovah's Witnesses disembark on Swans Island and conduct a profitable campaign of house to house selling of pamphlets and books. They themselves might be surprised at the number of those who purchase the books actually read and ponder them. The Pentacostal Church successfully missionized the island some years ago. The spiritual transports, the talking in tongues that mark the meetings of this sect make it the ultimate in emotional and imaginative appeal. At first the meetings were in a one-time boarding house perched bleakly above the remote Toothaker Cove. Later they moved to their own chapel near the village of Swans Island.

In the minds of many Americans, the New Englander is stereotyped as a cold fish—a shrewd dead-panned individual whose sole emotional outlets are whittling a stick and skinning a neighbor in a trade. This bit of folklore has been fostered by writers and humorists of New England for over a century, from Jack Downing to John Gould, but it still is preposterous. Hawthorne, Mary Wilkins Freeman, Eugene O'Neill have divined the wells of emotion and passion beneath the New Englander's dead-pan. The Yankee has a way of exploding into feeling after years of build-up of inner pressure. Emily Dickinson, as passionate a poet as ever lived, is the supreme example, but there are countless others. The list of New England reformers—Garrison, Whittier, Beecher, Margaret Fuller, the Peabody sisters, Horace Mann, Mary Baker Eddy, the Brook Farmers, the Shakers—is enough to refute the notion that New Englanders are emotional pygmies. With them the religious emotions are most sensitive, most prone to stormy upheaval, as might be expected in a region founded by religious zealots so fanatical as to give up friends and homes and goods to cross three thousand miles of ocean and settle in a wilderness inhabited solely by savages and wild animals. Smouldering deep in many a Yankee mind, no matter how uneducated, is a religious potentiality that a zealous preacher might puff into a blaze.

Men who live in isolated places think much of the eternal abstractions—God, life, time, and death. On the islands these thoughts when thought by intelligent men with a long religious tradition behind them are apt to be deeply absorbing. The problem of good and evil—a problem that a hundred years ago everyone in New England from the preachers and professors to the hired girls and farm hands in the kitchen discussed in deadly earnest—is urgently presented in the beauty and cruelty of the two-faced, treacherous sea. The speculative turn of mind is by no means lacking in men without formal education, particularly in these island descendants of the Puritans who are enough removed from the distractions of our mechanized society to have a little time to satisfy an innate bent for thinking. Few of them have failed to ponder that age-old question so clearly put by the sea: Why does a Creator who is at once all-powerful and benevolent mingle good with evil, beauty with ugliness, gentleness with viciousness, as He has in the ocean? Many islanders keenly feel the paradox though they do not couch it in terms of philosophy. A woman can see the wild beauty of the ocean on a northwest day, but knowing her husband is on those white waters she shrinks from the sight of them.

All island men have a deep respect for "the water," not only its storms but its fogs, currents, jagged reefs concealed like the claws of a great purring cat to tear the planks from any unwary boat. The malice of the sea, to them, is of many moods. Yet they see the splendor, too, even in its storms, when the waves seethe and crash and leap on the shoals among the islands with the noise of a million stonecrushers. Some men more than respect the water; they fear it with such a subconscious dread that they can't venture on it, even to cross the Bay, without deathly sea-sickness. But the same ocean that their whole organism revolts so violently against also insulates them from the terrors of the mainland, and they cling to their death-encircled island home as landsmen—farming, carpentering, storekeeping.

Only the summer folk on the islands are carefree around the water because they seldom see it in its wintry role and because the history of their families is not darkened by drownings in each generation. To the islanders the risks the summer people sometimes take seem foolhardy.

In both men and women the sea has inculcated a kind of fatalism. It will take its toll no matter what one does or says. No human attitude is simple and that of the islander toward his surroundings is particularly ambivalent. Celia Thaxter, the most articulate of all native islanders, expresses poignantly the fear and the iron attachment that holds these people to their sea-shrouded homes. She writes, "And all the pictures over which I dream are set in this framework of the sea, that sparkled and sang, or frowned and threatened, in all the ages that are gone as it does today, and will continue to smile and threaten when we who listen to it and love it and fear it now are dust and ashes in our turn." Elsewhere she states: "I believe, I am *afraid*, I can never put my heart into anything that doesn't belong to the sea." She realized that this state of mind was not always healthy. Apropos of a shipwreck in which two men had drowned, she wrote to a friend: "I am too much alone and get sadder than death with brooding over this riddle of life; and Nature is so placid; and the sea and rocks have ground out the life of those two to whom life was so sweet." If the theory that the stuff of poetry is conflict and tension, then island life is fraught with poetry of the most elemental sort.

Poetry and religion are closely akin and both are functions of the intuitions and feelings rather than of the reason. Hence the more emotional and revivalistic Protestantism of past centuries and of such present groups as the Pentecostals can best reach those whom reason has not aided the least jot in understanding the mystery of good and evil. The rational humanitarianism of some modern eastern Protestant clergymen, with their reduction of the Bible and the sacraments to myths, symbols, and memorials, fails to meet the need of these people. They require a feeling of direct communion with God and often achieve it by themselves. An extemporaneous prayer at a Ladies' Aid meeting resounds with sincerity and depth. One island-born woman who practises her profession of nursing on the mainland writes back to her mother inspirational poems which she states "are gifts from God." Her mother proudly passes them on in order to share God's bounty with others.

The older religions, like the evangelistic newer ones, offered not only emotional outlet but an understandable account of the world. The world was imperfect because of Adam. Evil was the devil's work and God tolerated the devil as a punishment of man. Man's way out of the dilemma was through belief in Christ, who offered salvation for all sinners who placed themselves in His hands. God and Christ saved one from the ever-present hell of this life and the world to come; they answered the cruelest of all enigmas—death. One need only study the epitaphs in the graveyards to understand how deep a comfort this faith could bring.

But there were payments to be made here on earth for this gift of comfort and salvation. The saved soul would show its blessedness through moral living. Morality in rural nineteenth-century America was a negative value—a series of *Thou shalt not*'s appended to those of the Decalogue. The most harped-on among these new commandments were the ones "Thou shalt not smoke" and "Thou shalt not drink alcohol." In the popular mind religion seemed to consist mainly of these two prohibitions. And there was some basis for this distorted impression. Maine itself had been prohibitionist since 1851. Through local option Swans Island has perpetuated the tradition to the present day. The most ardent and vocal dries are usually church members.

So closely connected are religion and teetotalism that a drinker wishing to reform usually signalizes his intention by joining a church. Sometimes his reform is permanent and wins the acclaim of the townspeople, who, whether church members or not, attribute the miracle largely to the influence of of religion. On the other hand a habitual drinker who might feel himself too weak to overcome the habit would hesitate to join the church. Thus church membership has suffered from this identification of religion with abstention from drink and to a lesser extent from tobacco and cards. Many an islander likes his cigar or pipe and many another his beer or rum. Their feeling towards church members who frown on these pleasures, no matter how well controlled, is that they are Pharisees who glory in trying to make everyone else as sour as they are.

As in all prohibitionist communities, Swans Island has always had its share of drinkers both moderate and heavy. Liquor can now be bought legally on the mainland and carried onto the island by the purchaser. For generations the smacks that periodically come to pick up the local lobster catches carried a liquid cargo for the fishermen. When fishing is good and money plentiful, the evening after a smack visits the island is apt to see some wild doings. Drinking is often excessive—the attempt to make the most of an opportunity that isn't always available in a dry community.

Yet full-fledged alcoholism is rare. Only by teetotal standards is drinking a serious problem. One island minister said that if he had time to explain in detail to every citizen that the Church and the WCTU are not the same organization, he might be able to fill his pews. He is probably over-optimistic because there are other obstacles both in the church and in the people to overcome, but doubtlessly the church would benefit greatly if in the public mind it were separated from the idea of spoil-sport self-righteousness in regard to even the most controlled indulgence in alcohol. However, the removal of these notions will not be easy. Once it conceives an opinion, the island mind is loath to give it up. One minister explained that some islanders do not go to church because of something another minister said twenty or fifty years ago.

That the Church exists at all on Swans and some of the other islands is due almost entirely to the efforts of the Maine Sea Coast Msision. This non-denominational society was started at the beginning of the century "to undertake religious and benevolent work in the neglected communities and among the isolated families along the coast and on the islands of Maine, and to engage in all efforts that are calculated to contribute to the moral and spiritual welfare of the people in question." Its territory extends from Kittery to Calais and includes ten thousand people. Though some islands like Swans had their own churches and regular pastors, others remained almost as completely cut off from their spiritual heritage as they were a century before when, for example, a minister visiting strongly devout Monhegan, where most religious facilities were lacking, would find a backlog of weddings to perform and funeral services to conduct over filled graves.

A succession of Mission boats, the last three of which have been named the *Sunbeam*, have made the rounds of this scattered "parish," carrying spiritual and medical aid, removing sick people to mainland hospitals, breaking ice in winter-locked harbors, bringing Christmas presents to every child, old person, and shut-in on the islands and in the lighthouses. On several of the islands the Mission maintains a full-time minister and aids the town or plantation in finding and supporting a nurse or physician. It helps find suitable teachers for the schools—teachers with an ideal of service rather than derelicts who can get a job nowhere else and have to accept low island salaries for lack of anything better. It conducts dental clinics for a number of communities and pays for tonsillectomies for needy children. Though its policy is against owning church property, it occasionally rents or buys and sees to the upkeep of a parsonage, a church, or a library.

The Mission is supported by voluntary contributions, many from people who spend their summers on the Maine coast. Among the largest donors is Sigma Kappa Sorority, which has long made the Mission one of its philanthropic projects. Yearly it provides enough money to pay part of the salaries of two full-time mission workers and it supplies a large number of the Christmas presents the Mission distributes. The benefiting communities also contribute small amounts to the Mission, but the tradition of receiving has become stronger than that of giving. The total Swans Island contribution in 1953 was seven hundred and seventy dollars. During the depression it was about two hundred dollars annually. A minister who watched the island in one year invest many thousands of dollars in TV sets views this situation with a jaundiced eye.

Over the years the Mission has spent thirty thousand dollars on Swans Island—more than in any other community. It provides a minister who serves both the Baptist and the Methodist churches, paying him a salary of twenty-four hundred dollars a year and providing his house. Several able and devoted men have held this position, struggling to build up the congregations, encouraging activities for the young. But their labor has brought disheartening results. They have kept church membership from vanishing altogether; they have done much counseling in personal and family problems. Through the Scout troops, the presentation of movies, and Sunday Schools they have helped put the brakes on the unruliness of adolescence. But no community awakening, moral or spiritual, has taken place. These pastors, with whom the Adventist minister has cooperated closely, have been fighting a holding operation. They have gained no appreciable ground, but also they have lost none.

The Mission's policy is to make its projects self-sustaining as soon as possible. The communities are expected to fend for themselves as much as they can. Yet on some of the islands the Mission and the summer people have done so much that the people have developed a supine attitude of letting others shoulder their burdens. On Swans Island some of the summer people who contribute to the Mission are poorer financially than many of the islanders who give nothing. The assumption that the islanders are incapable of looking out for themselves is so deeply ingrained in these summer people that they continue to give when the need no longer exists, and the islanders, probably blind themselves to the changing pattern, persist in being the beneficiaries. Yet they retain the touchiness of the truly needy and doubtlessly feel quite sorry for themselves because they are the recipients of "charity." The ministers and other Mission workers are aware of their attitude and try to combat it by getting the people to think and live in terms of giving rather than receiving. Thus in 1953 the Swans Island pastor was justifiably proud when he was able to collect from the islanders one hundred dollars to put two youths through high school in India. He was quick to point out the significance of this: The remote island hamlet forming ties with the peoples of one of the world's most populous nations.

2

When the island is without a doctor the Mission has helped recruit and support a public health nurse. The following agreement illustrates the way in which the town and the Mission cooperate.

NURSE'S AGREEMENT

Agreement between the Town of Swans Island and the Mission, voted on at the Swans Island Town Meeting March 4, 1946, concerning the employment of a Public Health Nurse:

The Town of Swans Island will pay the entire salary of $1,800. They will also contribute the sum of $75 for the 1946 dental clinic. The Town also voted to raise and appropriate the sum of $50 to help in the payment of off-the-island telephone calls made by the nurse in line of duty and not paid for by the sick person or his family for which calls have been made.

The nurse will charge a fee of fifty cents for a daytime call, $1.00 for a night call, and $2.00 for an all-night case. Monies thus received will be turned in to the Town treasury and will go toward paying the nurse's salary.

The Mission guarantees to pay a maximum of $35 a month car allowance. The nurse will charge fifty cents for use of her car when called out on a case, in addition to the nursing fees referred to above. At the end of each calendar month the nurse will credit the Mission with the total amount thus received toward the use of her car and the Mission will be liable only for the balance to make up a total of $35.

In consideration of the payment of the bulk of the car expense by the Mission, the nurse will be free to carry on her program of preventive medicine and public health in addition to caring for the sick.

Currently the Mission is trying to establish at Old Harbor a resident doctor and a clinic to serve Swans and Long Islands. Because no adequate house at reasonable cost is available, fully modernized prefabricated buildings would be put up for the doctor's dwelling and the clinic. The latter, in addition to its up-to-date medical equipment, would have several rooms for temporary and emergency hospital cases. For this project the Mission would raise up to forty thousand dollars. The island's share of ten thousand dollars could be paid by each fisherman's contributing for two years the proceeds from one lobster trap. Special town meetings were held to discuss the plan, committees were appointed, but nothing has been done. The taxpayers are still quivering at the expense of the new school. The courage and leadership are lacking to go on with the "hospital," as the islanders insist on calling it. The crepe-hangers are busily at work, but soon their day will pass and a timid step forward will be taken. Then one day the landing scow will arrive with the prefabricated houses and the clinic will be a reality. The Mission has learned that "it takes six years to get anything started in a Maine community. For two years everybody is agin' it. For two more years they think it may be a good idea, but they couldn't do it; at the end of two more they decide it might not do any harm to try it."

The great contribution of the Mission to communities like Swans Island has been in bringing nurses, doctors, and ministers of the highest professional and personal quality into them. What these men and women have accomplished cannot be measured by numbers attending church or receiving dental care or having tonsils removed. Their most valuable service has been their mere presence in the communities—their intangible influence as friendly, sympathetic, able human beings devoted to unselfish service. Most of them have entered deeply into the life of the town. One minister was elected moderator of Town Meeting for several years. When they depart, as they all do after a few years, they leave lifelong friends behind them. Their memory lingers and they become part of the spiritual tradition of the island and their virtues become magnified even beyond what they actually were.

To take an example, few people will forget the "public health nurse," Mrs. Osborne, who was on the island with her two daughters during World War II while her husband served as a commando in the British army. She had been one of Dr. Grenfell's nurses in Newfoundland and Labrador. Much of her married life she had spent on an uninhabited island in the Straits of Belle Isle, where her husband kept a lighthouse which was visited once or twice a year by a government vessel. With her energy, her intelligence, her humor, and her experience she made an impression on Swans Island which yearly grows stronger with those who knew her.

Some of the physicians who have been on the island have also left their mark. The first ones came and established their practices like any doctor on the mainland. Later the town guaranteed them a minimum income or paid them a bonus above what they made in fees—in recent years, twenty-four hundred dollars. Often the doctors have been prominent in town affairs. Several, as the most highly educated men in the community, have logically been chosen as chairmen of the school board. One early physician, Dr. Small, wrote the genealogical history of the town in 1898, and served not only as superintendent of schools but as town clerk. A recent physician was a leader in getting electricity on the island.

The worst drawback to practicing medicine on the island is that often there is not enough to do. Many islanders believe in doctoring themselves, in "toughing it out" even if they get seriously ill. They have a few favorite nostrums dating from the last century: creosote for coughs, checkerberry for stomach disorder, paregoric for anything at all. Perhaps the tradition of self-doctoring goes back to the days when there was no physician and one had to rely on his own knowledge of herbs or sail to the mainland for professional help. Also, recent doctors have found that as a group, islanders enjoy remarkably good health. One complained: "Everyone here dies of old age. There is nothing to interest an M.D." Dentists brought on by the Mission clinics comment on the good condition of the islanders' teeth, though there is no indication that an unusual amount of toothpaste is sold or used there.

The island provides a challenge for the doctor or nurse more interested in doing a good professional job than in making a fortune. In an emergency, other help is far away. The doctor or nurse is entirely on his own with no pulmotors or asceptic operating rooms, though recently an X-ray apparatus has been purchased by the Mission. A story that has become part of the lore of the island casts a heroic light on one of its best-loved doctors. During a northeast blizzard—raging as only a blizzard on the coast can when they "get a good one"—the Baptist minister was taken with acute appendicitis. Immediate operation was a must to save his life, but there was no way of getting him to the hospital across the wind-stewed cauldron of the Bay. The doctor decided to put him under the knife at once. A graduate nurse was living on the island at that time. A neighbor undertook to get her with horse and sleigh at her home on the remote east shore of the island, though the town roads were so exposed to the gale that they were drifted too deep even for such travel. By following a network of wood roads in the interior the driver reached the nurse's home and brought her back to the parsonage. A table was improvised, oil lamps, reflectors, and flashlights were arranged to throw the most light. Together the doctor and the nurse successfully administered the anaesthetic and performed the appendectomy.

Whenever possible the seriously ill are rushed to a hospital. The *Sunbeam* is always ready for such work. Built as an ice-breaker, she is sometimes (as during World War II) the only means of getting off in severe winter weather. Many an islander remembers the *Sunbeam* with deep thankfulness. In recent years most women have gone off the island to have their babies. Still there are occasions in illness, accident, or child-birth when there is no time to get ashore. Then the minister, the doctor, the nurse, show what they are made of. Usually they measure up.

11

Schools

IN JANUARY, 1954, THE CHILDREN OF SWANS ISLAND, MANY OF whom had never used a flush toilet, had the exciting experience of attending a school with modern plumbing and central heating. The building is also equipped with a kitchen from which it is hoped hot lunches will be served with the usual state and federal aid. For a few of the children these lunches will be the first full meals they have ever sat down to and eaten all at one time.

The history of the island has been one of gradual drawing together of widely separated groups and individuals—a process in the growth of a community much more important than mere increase or decrease in population. At first there were four main sections on the island, connected only by footpaths and ox-roads through marshes and over rocks. In those days the easiest communications were by water. Later with the coming of horses the roads were somewhat improved and there was more intermingling. The bridges of granite blocks and boulders were built for the ages, and the energetic hiker can still find them deep in the spruce thickets.

About 1890, a telephone wire was strung between the stores on opposite sides of the island. Soon houses in between had their lines, and branches radiated out to either side. A central office was installed, the telephone company was incorporated, and a cable was laid to the mainland. Before long one hundred houses, twelve to fourteen on a line, were served. Old residents consider the telephone to be the greatest single convenience ever introduced to the island.

At the turn of the century the three villages which made up the island (the North was almost completely deserted) were still far apart socially, and they later relinquished their individuality only with reluctance. Then tons of dynamite were brought on to prepare the roads for the advent of the automobile. Motorboats, too, were making travel from one side of the island to the other easier. Unification of the township was steadily progressing on the geographic level. But the school committee and the board of selectmen were, and still are, made up of one member from each of the villages; each "side" has kept its own church and post office, for R.F.D. is looked upon with horror.

Civic organization as well as roads and telephones furthered the consolidation of the community. In the early days families in the different parts of the island chipped in to hire teachers, whom they paid nine shillings a week ($2.25). Classes were held in private dwellings. In 1834 the island, its population grown to over two hundred, took the first political step. It formed itself into a plantation and the first school house was erected near the Carrying Place. A young woman from the nearby coast was hired to teach its first term. No records of the appropriations made in the first plantation meeting exist, but four years later, in 1839, ninety dollars were allotted for the school and one hundred and fifty dollars for roads. Among the earliest teachers was a lad who later became a judge of the Maine Supreme Court.

In these times two terms of school of about three months each were held—one in the summer and one in the winter. The average wage for a teacher in the winter term was fourteen dollars a month plus room and board, which the town paid some islander to supply, sometimes for as little as a dollar a month.

Summer teachers' wages were a dollar a week. By 1842 there were four districts on the island and by 1859, six districts.

Slowly the school appropriations mounted, but the road appropriations held the lead. In 1850 the four district schools then in existence had one hundred and ninety-four "scholars" —an indication of the size of families one hundred years ago, for the total population was no larger than at present, when the enrollment is about sixty. One of these one-room schools had sixty-four scholars, so that the personal touch between the teacher and the handful of students that sentimentalists suppose to be typical of such schools was here desperately lacking. The most frequent touch must have been that of the teacher's birch upon the buttocks of rowdyish boys. In few schools of the present day is one teacher afflicted with the complete supervision of so many children and never of such diversified age. In this year the plantation earmarked one hundred and sixty dollars for education. The schooling of each of the one hundred and ninty-four students for an entire year cost eighty-two and a half cents.

In 1893 the school appropriations reached six hundred and fifty dollars. Women teachers were still being paid as little as two dollars a week. Often they gave up their classrooms to work in the lobster canneries, where they could make three dollars and fifty cents a week. Until well into the present century the teacher on nearby Calf Island, now deserted, received only four dollars a week, though with only five or six pupils the per capita expenditure was above that for the rest of the town. By this time the annual commitment for roads ran two hundred dollars behind that for schools.

The following year the state began extensive support of public education. Immediately the Swans Island appropriation dropped to five hundred and five dollars, a sum approximately equal to the state aid. Teachers were now making around eight dollars a week for a school year of twenty-six weeks. Some teachers were able to increase their earnings as much as fifteen dollars annually by being janitors of the buildings in which they taught. Under the stimulus of the new law and with greatly improved roads, the first move towards centralization was made. As the community was drawn together, the number of schools diminished. Three districts were consolidated, a single two-story building being erected for the purpose.

When the island became a town in 1896, a more self-sufficient political unit than a plantation, advances in education were numerous. Within a few years two other two-room school houses were constructed. The best and last of these cost two thousand dollars, of which two hundred dollars was raised by the town and eighteen hundred dollars "hired." One of the teachers in this school received two hundred and twenty-eight dollars for twenty-six weeks of instruction—or slightly under nine dollars a week. To aid him, perhaps, in his struggle for existence, the citizens elected him Town Clerk, for he was a native islander, and in this capacity he augmented his income by eighteen dollars and fifteen cents a year. Not long afterwards he left, as have many good island teachers, for a more attractive job on the mainland. Eventually he became principal of a high school in one of New England's largest cities.

From the turn of the century to the end of the First World War the town allocations for education rose to one thousand dollars. Most of this money went towards teachers' salaries and was more than equaled by state grants, which in the same period rose from six hundred and ten dollars to fifteen hundred dollars. Other sums, varying from year to year, were voted for upkeep of buildings, text-books, and other supplies, and tuition for students who had to go out of town for high school, since there was none on the island. Education was the biggest item in the town budget, though running a close second were roads, for which state aid had jumped from zero to $1,925 in 1916. Six teachers were employed, two for each of the island's three schools. But teachers' pay was at a standstill. Just before World War I the teacher of the upper grades —the "grammar school," as it was called—received $12.77 a week for a thirty-week school year. The teacher of the lower grades, the "primary school," received $8.26 weekly. Even in 1919, during the post-war teacher shortage, the average yearly wage was about five hundred dollars.

The war winter of 1917-1918 was a superintendent's nightmare. The school board "drafted" some teachers from among the island women. For a time in that winter the schools were closed altogether by the flu epidemic. Some of the teachers who came from away resigned at the end of the fall term on account of the epidemic and severity of the weather, which was the coldest in history. Whether the low pay had anything to do with their leaving, the superintendent did not say. One replacement eventually was located "to the westward," but she got to Rockland only to find the steamboat ice-locked in the harbor. She lingered for a day or two and then went home. The school room she was destined for had no teacher for six more weeks. The only other case of such lengthy discontinuance of classes was in 1902 when one school was closed for almost a term because the teacher hired for it proved unable to keep discipline.

Until 1910 the superintendent of schools was a member of the local school board. His pay was from fifty dollars to one hundred dollars. For many years the town historian and physician, Dr. H. W. Small, had held the position. In 1911, the town voted to share a superintendent with Vinal Haven and other nearby towns. Later Swans Island was under a superintendent residing at Southwest Harbor and at present is under one with headquarters at Stonington. Swans Island's share of the superintendent's salary has risen from one hundred dollars to four hundred dollars. This official, who visits the island at least once a month, draws up the budget for the schools, makes plans for repairs and new buildings, and supervises the teaching personnel—all of course in council with the school board and eventually with the voters in town meeting.

Between World War I and the Depression, the salaries of the five or six teachers on the island rose to six hundred and sixty dollars a year. Town allotments for education had risen from twelve hundred dollars to four thousand dollars, though state aid continued to hover around fifteen hundred dollars. By 1921 the yearly expenditure for each child was fifty-five dollars as compared with approximately two hundred dollars on each mile of the town's roads. During the Depression the salaries of teachers dropped back to four hundred and eighty dollars; the town appropriations to two thousand dollars, including eight hundred dollars for tuition of students attending high school off the island; and the expenditure per pupil averaged thirty-five dollars. Roads were being supported at the rate of one hundred and ninety dollars a mile. Lobsters, the economic mainstay of the island, were selling for twelve cents a pound and were scarce to the point of non-existence. At one time there were sixty-six delinquent tax-payers; the poor rolls were fuller than ever before.

During World War II and the post-war inflation, lobsters and education made a comeback. For the year from town meeting in March, 1952, to town meeting in March, 1953, the citizens voted $8,670 and the state supplied $3,462 for the schools. Two teachers, both islanders, were receiving nineteen hundred dollars each. In addition eight students away at high school were costing the town $369 each. That fall the schools were centralized and a surplus of almost four thousand dollars was laid by against the cost of a modern building. A bus was purchased to transport the children to the central building, though buying and running it slightly exceeded the saving in salary for the one teacher who was eliminated by the arrangement. The per capita cost of educating the children was $104 while roads cost $582 a mile.

2

For at least fifty years the town has sporadically discussed having a high school. In 1906 and 1907 there was a ninth grade, in the first term of which forty students enrolled. But the large initial attendance was prompted more by curiosity than by desire for learning. In the second term lack of interest on the part of parents and children caused a sharp drop. The next year the superintendent recommended giving up the high school and concentrating on the improvement of the lower schools. Many Swans Islanders have always held a poor opinion of formal education, and wish neither to support a high school on the island nor to send their children to one elsewhere. Even today the comment is occasionally made that school is for girls, "who don't know any better than to study." Many young people drop out of school as soon as they can legally and talk about going to the mainland to become mechanics. Older men who have had the same dream and found almost all jobs closed to them because of their lack of a high school diploma look sourly on the ambitions of these young upstarts and warn them to get more education or drop their "big ideas." Needless to say, the warnings make no impression.

The lack of an island high school is felt in many ways. Youngsters just out of elementary school are not old enough to settle down to regular lobstering or other work. These footloose teen-agers lounge about the island, their thoughts inevitably bent on alcohol and sex. If they can acquire a jalopy in which to hurtle along the roads with their girls, their bottle, and other admiring striplings, they are in paradise indeed.

In assembling such a vehicle the theft of a neighbor's battery or tire is occasionally a necessity that is not allowed to thwart their plans. Night sessions in parked cars are held among teen-agers, and parental shotguns are sometimes, figuratively, brought into play in the establishment of new families. Saddled with a wife and child, with other children soon to follow, the boy must abandon his dreams of becoming a mechanic on the mainland, and he must get a boat and start lobstering. Many a family thus created has become a mainstay of civic and moral virtues, for no disgrace is connected with such marriages. But also more than one unhappy marriage, marred by drunkenness and infidelity and terminating in divorce, has resulted.

Advocates of extra-curricular activities in our schools would find on Swans Island much support for their position. For here the schools sponsor no organized athletics; it is perhaps the only community in the United States where the boys care little about playing baseball. From time to time music is taught and encouraged, most recently by the wife of one of the island ministers, but such ventures soon are discontinued. Yet there is a flare for music among the islanders, and the many who have given it a try sing and play with great pleasure to themselves and others. One musical family often performs as a group in the churches. But fewer and fewer island children study an instrument. School dramatics and newspapers are unheard of, and folk-dancing, or dancing of any kind, is almost completely neglected. When island children gather for one of their infrequent "socials" at a community hall, bedlam ensues. The children race about the room shouting and consuming untold bottles of "tonics," the routine being broken by occasional trips of boys and girls to the curtained jalopies or alder thickets outside.

In a place so highly endowed by nature, one might expect that the out-of-doors would lure the youth into all sorts of wholesome pastimes. To be sure, the young people are ardent skaters, and formerly, when the winters were severe enough to freeze the coves, ice-boating provided much sport. The toddlers paddle about on the beaches, and when they are a little older row their fathers' punts around the wharves. Every boy becomes an expert oarsman by the time he reaches adolescence. Either "pushing" or "pulling," the lads swarm about a harbor with skill and speed that are the admiration of a landsman. Also, in recent years the young people have done some swimming entirely on their own, in which way they have an advantage over their elders, most of whom wistfully admit they "swim like a hunk of iron, straight to the bottom."

But interest in nature and the out-of-doors vanishes with the onset of puberty. Many island adults cannot name the commonest plants and birds. They refer to white alders as "those bushes with the red berries, whatever you call them," or to the shellfish known as a sand dollar as "those round things you find in the sand." If a boy thinks of hunting, he thinks of "jacking" deer, a practice punishable by a fine of two hundred dollars. Or he considers himself lucky if he can overtake a swimming deer and, from his boat, brain it as it flounders helplessly in the water. Fishing, to the youth, is a form of labor indulged in by his elders and to be avoided for as long as possible. Camping, hiking, nature study, he dismisses as sissy stuff.

Many people think that the only thing needed is leadership. The young are bright, healthy, remarkably "handsome." To see a group of these fair-haired, blue-eyed Anglo-Saxons dressed and polished for Sunday School is a reassurance that the human animal can be physically attractive. The good looks and healthiness persist through adolescence. Acne is not a scourge of the island young. There is nothing wrong with the material. What is wrong is the neglect and mishandling of it.

Recently a minister and his wife, who was the island physician, organized Scout troops for the boys and girls. For the first time the children had the experience of sleeping and cooking their meals out, of working for achievement badges. The birds, animals, plants and trees of their island were called to their attention and named for them. In hikes along the shores and on the abandoned roads they saw the forests, coves, escarpments that draw summer people from hundreds of miles away. Visiting deserted localities and searching old cellar holes, bridges, and quarries, they learned something of the history of the scrap of land of their nativity. Occasionally some summer person who puts in a long season contributes a knowledge of ornithology, botany, or dramatics.

But on Swans Island efforts have not yet been prolonged enough to bring about decisive results. The leaders, who are almost always ministers, teachers, doctors, nurses, or summer people, move away just as pay dirt is reached. The parents continuously remark that they want their children to have all the advantages those on the mainland have. Yet they do not seem aware that parents on the mainland bring these advantages to their children by hard work as Scoutmasters, PTA members, and plain citizens.

An example of what prolonged, devoted, and intelligent work can do with the children of an island community is the work of Mrs. Gladys Muir at Frenchboro on Long Island, just four miles east of Swans. For more than twenty years she has served this village of about eighty people as teacher under the pay of the Plantation, and as minister under the pay of the Maine Sea Coast Mission. Frenchboro, until recently without electricity, and with only three-quarters of a mile of road, is one of the most remote communities on the Atlantic coast. For generations, as a port where off-shore fishing vessels would put in to provision and give the crew a chance for a "good time," Frenchboro was only semi-civilized. The change had begun long before Mrs. Muir's time, but she has done much to make it a better community. Hers is one of the nation's model one-room schools. Her methods are progressive without being chaotic. In her teaching she uses the materials at hand. Thus she studied bird and marine life at the Audubon Camp at Miscongus Bay, Maine. With the knowledge thus gained she introduced the Frenchboro children to the world surrounding them. As a well-known ornithologist pointed out, these children trample on interesting and significant natural things every time they walk down to the beach or take a stroll through the woods. All they need is someone to point them out and they will be as well off as if they were in an expensive biological laboratory. Revealing to people the meaning of their surroundings is a rare art. Yet that is what this teacher does. The school is extended out-of-doors onto the beach at the foot of the hill and into the woods that creep down to the edge of the playground.

Since Frenchboro is completely Protestant, Mrs. Muir and the school board have received special permission from the state to combine religious teaching with the regular school subjects. The opportunity is unique, not practical in most places, but it is an example of what can be done under the unusual conditions of some of the island communities. The children on all the islands are responsive to those who sincerely wish to help them. There is nothing sullen and vindictive about them, nor are they prematurely sophisticated. They take new pupils into their midst with the utmost friendliness. One twelve-year-old girl after her first day in the island school, which she had entered in the middle of a term, said: "These are swell kids. They're friendly and nice."

3

Still, whatever the schools can offer the children on Swans and the other islands will have to be on an elementary level.

The islands are financially unable to support a secondary school, and many islanders have no complaints. But many others have always wanted more education than the town could give them. Fifty or sixty years ago, after the new school buildings were completed, the "upper room" teachers—the higher grades were appropriately held in the upstairs rooms—would often offer advanced work for students interested in taking it. In this way ambitious young people could pick up algebra and language equivalent to a mainland schooling through the tenth grade. Some of the older inhabitants still speak gratefully of these teachers who took the trouble to give them a chance otherwise denied unless they went to the mainland. They speak with particular affection of a Mr. Williams, a Bowdoin graduate, who to their minds "seemed to know everything."

But in general children wanting a secondary education have had to go off the island to get it. The town and the state contribute towards tuition and board and room for any student who lives in a town without a high school but wishes to go to one elsewhere. In 1953 eleven students availed themselves of the opportunity. Fifty years ago not more than four or five did so in any one year, though the population of children was twice that of the present. The number going beyond the elementary level has risen from one or two from each graduating class to an entire class of five a few years ago. Lurking in everyone's mind is the fear that someday the children will have to live off the island. "If the fishing gave out here, what kind of work would a lot of these young fellows find on the mainland? How far will they get without a high school diploma?"

There is one big advantage in their having to go away to school. The pupil can choose his school from all those accredited by the state. Some students go to schools in towns where they have relatives with whom they can board cheaply, for the state and town allotment doesn't cover all expenses. Also, transportation facilities to and from the island often determine the town to which the children go. While the steamboat was making daily runs to Rockland many island children went there or to Stonington, the first stop along the way. But many, in search of quality rather than convenience, have gone farther afield. Higgins Classical Institute, a first-class Maine academy, has always drawn a high percentage of island pupils.

Attendance at high school might be greater except for the disabling homesickness that overwhelms some of the children as soon as they leave the island. The complexity of mainland life is too much for them. They yearn desperately for the familiar wharves and boats, spruces and rocks, and peak-roofed houses cuddling against the hillsides. On his island the child seldom has had to cope with a person he doesn't address by first name. Furthermore, he can be sure that everyone knows all about him and takes an interest in his affairs. Then one day this child of thirteen or fourteen steps out on a mainland wharf, boards a train or bus, and is hurtled away to a strange town which, though it may number less than a thousand, is a metropolis in comparison with the up-cove village of his birth. Everywhere are strange noises and sights—factories, solid blocks of brick buildings, inconceivable motor traffic, and gaudy store fronts which scream out the stridencies of juke boxes and radios. And everywhere are strange people, strange faces. The waitress at the soda fountain doesn't give him the clipped "Hi" plus his nickname that at home went unnoticed. The pedestrians on the sidewalk fail to greet him with the islander's nod and dry jab of humor: "Did you get up before breakfast this morning, Jim?" Bewildered, the child may flee back to the island the next day or the next week. Or, overreaching himself in his efforts to make good, he may fall in with unruly companions, outdo them in wildness, and sometimes end in jail or in a drunken accident.

Sometimes a summer person will attempt to give a particularly promising island child "a chance in life," this being to transplant him into some vast city where he is put into high school or college. In his novel *The Islands* Gerald Warner Brace has made an imaginative study of just such a case—an orphaned island boy dragged off to Boston by a well-meaning spinster rusticator. Surviving the rigors of life on Beacon Hill for ten years, the boy eventually throws over a chance to study engineering and returns to his island home to build boats with his uncle.

Of course most of the children make satisfactory adjustments when they leave the islands. After a period of moping, they get down to their studies, find friends and sweethearts, and are kept happy by periodic visits home. Some parents believe that it's good for them to have this experience, "to see something else," and this is an argument used against having a high school on the island.

In recent years a number of island boys and girls have gone on to college. Usually they enroll in one of the state normal schools or in the University. The college graduates don't often return, though more than one B.S. or A.B. is hauling traps along the coast because he finds lobstering to be more lucrative than the teaching he studied for. Most of those who do make a life for themselves elsewhere are leaders in their communities and their professions. As one clergyman who has spent a life-time on the coast said: "The chief contribution of these islands to the rest of the state and nation is human beings. And they export a fine product, with an emphasis on quality rather than quantity."

Most of the high-school graduates return home to become lobstermen or the wives of lobstermen. For them the pull of the islands is too strong to break permanently. Thus there is little turnover in population. The same families persist from generation to generation and year after year the birth and death rates miraculously balance each other. But the hard-earned high-school education is not wasted. Those who have managed to get their diplomas become selectmen, school committee members, or officers in some community enterprise like the electrical co-op.

Various schemes have been proposed for making high school education more accessible to Swans Island children. In 1950 the town of Brooklin on the mainland directly to the north offered to take all Swans Island students wishing to attend high school for the yearly sum of ninety dollars each, which is considerably less than the tuition charged elsewhere. The Brooklin school is housed in a modern building with kitchen and gymnasium and has an enrollment of seventeen students with two teachers; in 1955 it had its first graduate. The school committee thought the school could be run much more effectively, not to mention more cheaply for the Brooklin taxpayers, if the Swans Islanders would agree to the plan. In addition to low tuition rates, teachers of high quality and "a complete and enjoyable program of activities during and after school" were promised. Then, too, Brooklin is nearby. The island children could return home on week-ends. Also they would comprise perhaps "one-third of the school, which would mean that they would not be neglected as they might be in a larger school." This last point was a telling one. Island children in the schools near the coast have sometimes been looked down on as being the coastal equivalent of hill-billies. Their accent is different and they are unused to certain amenities like pool rooms, soda fountains, and bars. Their tendency to become either mopish or unruly through homesickness has only worsened their reputation. In a small school where they were a sizable block of students they might be more readily accepted.

But there was one serious drawback to the Brooklin plan. If adopted it would mean the town of Swans Island and the state would pay board and tuition only for students attending the Brooklin school. A choice would no longer exsit. Since many families wished their children to attend the older up-country academies, the plan come to nothing.

4

Thus the island's educational efforts have gone mainly into the improvement on an elementary level. When in 1953 the schools were consolidated the students were herded by bus to a sixty-year-old building, its paintless walls as gray as the ledge upon which it gauntly perched. It had two rooms laid on top of one another, with a privy attached by a long corridor like the tail of a fish. The "upstairs" overlooked the ocean and islands to the southwest as far as the rampart of Isle au Haut. Wood stoves with loose pipe zigzagging to the chimney threw out a wilting heat. The scabrous walls and floor were encrusted in dirt and grease from years of student lunches, and the corridor leading from the privy funneled out the characteristic stench.

The new school is all on one floor, has three rooms as light as the out-of-doors, central heat, and kitchen. To build the school the town had to put up six thousand dollars cash and borrow thirty thousand dollars through the Maine School Building Authority. The debt will be paid off at the rate of $1,320 plus interest a year. For this purpose eighteen hundred dollars will come from a tax always assessed on the towns by the state but in the last few years turned back to them to be spent for some local project. Many people think the commitment is excessive and that taxes will soar. Yet contracts were signed and scows began to land well-drilling equipment and building material.

There were misgivings a few years ago when the bus was

brought on. The expense would be too great, the children would get sick from riding on it. But the twenty-four hundred dollars spent in buying the second-hand bus and operating it for one year was more than offset by the saving in building maintenance and the one teacher no longer needed. The children loved riding the bus, too, and would sometimes get into it on the outward run so that they would have a longer drive. Adults, also, began to ride it from one part of the island to another. For the first time in its history the three villages, once quite isolated from one another, were connected by daily bus service—and free, at that!

Centralization has done much to give the island a sense of common purpose. When first brought to the same school the children from each village formed a group which eyed the others awkwardly, if not hostilely. The communities were divided not only by land and woods but by a one hundred and fifty-year-old sense of identity. The barrier between the groups of children was the same as that between the adults of the three localities. But ever since road improvement started fifty years ago the barriers have been slowly undermined. The consolidation of the schools may send them crashing. Already the children mingle without rivalry or shyness. The parents will be brought together more and more at school functions. In the building of the school there have been unusual examples of cooperation in the manner of an old-time barn or church raising. The men formed volunteer work groups to dig drainage ditches, lay pipe, and landscape the yard. The electric stove needed to complete the kitchen was not covered in the original appropriation; so a town-wide collection was made to purchase it. The children, also, have helped by giving card parties and numerous of their boisterous socials to raise money for a screen and black-out curtains to go with their newly acquired movie projector. The tax collector did her part by contributing her percentage of the collection to the school fund.

The chief problem now will be the securing of teachers, for the low salaries lure fewer and fewer from the mainland. Island women who are trained to teach but have become housewives have long been the mainstay of the schools. But recently they too have rebelled against the poor pay. During the year the school moved, two teachers from away—a man and his wife—were retained with difficulty at twenty-six hundred dollars a year each, six hundred dollars more than the native island teachers received. Later nineteen or twenty sub-primary children were assigned to the teacher of grades one to three. Few teachers will tolerate such a load for long. Yet the taxpayers, already cowed by the cost of the building, balk at hiring the much-needed third teacher whose room now stands empty.

Eventually there will be an adequate staff. The community, slowly realizing its oneness through roads and telephones and civic organization, has covered a lot of ground since the first teacher, for a dollar a week, taught forty or fifty students in the cabin that was the island's first schoolhouse. There is no reason to think the trend will reverse itself—unless the lobsters cease to crawl, and that just can't happen, as one school board member put it, with this new building to pay for.

12

Home and Social Life

ESTABLISHING A HOME AND FAMILY IS RELATIVELY EASY FOR young people who wish to remain on the island. For material possessions the couple need only a boat and a home. The young man can get a boat from the proceeds of a summer's work with a seiner or a year's partnership with another fisherman. His gear he can get on credit from a lobster buyer. He can purchase a house for one thousand dollars or rent one for ten or twelve dollars a month. Furthermore, there is no problem of adjustment to a new way of life. If the couple remain on the island, they will live within a few miles of their families and will continue all their old friendships and old activities. Thus, unhindered by many of the obstacles to marriage in the urban and more prosperous agricultural sections of our society, Swans Island couples usually marry young—in their late teens or early twenties.

The weddings on Swans Island are often boisterous. Shivarees, horseplay of all sorts, and drinking and noisemaking are part of the occasion. This merriment is the spontaneous expression of the young. It is severely frowned upon by many of the old people and is discouraged by the ministers, whose performance of the ceremony has sometimes been interrupted by a practical joke.

After the marriage the couple keep up a close relationship with their parents. Though they almost invariably have a house of their own, they try to reside in the same section of the island as at least one of the parents, whose wisdom and experience they value to an extent unusual in modern America. The woman is constantly and willingly under the tutelage of both her mother and her mother-in-law in matters of housekeeping. In regard to fishing the man frequently seeks the advice of his father, who has taught him whatever he already knows. This deference is not astonishing when one considers that fishing is a traditional skill to be learned only from those who have already mastered it. Long after marriage the young man must rely on his father—or less frequently his father-in-law—for unwritten information and lore that other fishermen would be reluctant to pass on to a competitor not related by blood. Also, there are ways in which father and son, though operating separate boats, may be helpful to one another. They can share a wharf or shore frontage; they can pick up bait for each other if they run across a good supply; they can team up for jobs that one man could not do alone, such as hauling up a boat or placing a mooring. Thus the strong tie between generations is one of expediency as well as sentiment.

On the other hand, during childhood and adolescence parental authority is usually rather lax. The girl helps in the kitchen and the boy goes in his father's boat as companion rather than mere helper. Much of the time the children are free to play, and as soon as they are in their teens they wander unhampered among the three villages, for one can almost invariably catch a ride in any car that comes along. Quite a few drive the family car at the age of twelve.

If a Maine island family had to choose between owning a good home or a good workboat, they would choose the boat unhesitatingly. With the boat they might eventually earn enough to buy a good house, which ranks second in the scale of material values. Yet the houses on Swans Island are sturdy and roomy and bespeak a firm tradition of family living. The old custom was for married people to have many children and to keep with them aged parents or grandparents. The families are getting smaller, though not so small as among apartment dwellers, and the old folks are still cared for to the end. Only four out of the one hundred and fifty-nine occupied houses have "1.51 or more persons per room," the ratio chosen by the Census Bureau to indicate where overcrowding begins. In some houses occupied by single persons or elderly couples the ratio runs as high as ten or twelve rooms to one person. Of the one hundred and fifty-nine houses, one hundred and twenty-five are owner occupied and thirty-four are rented. The proportion of rented houses and the average of fourteen dollars a month paid for them are normal for Maine rural and coastal villages. But the number of owner-occupied houses mortgaged—one out of twenty-five—is extremely low.

The first island homes were log cabins whose chinks were plastered with mortar made of clam shells from the heaps left by the Indians. The first frame houses came soon after, and a number of them are still lived in. They are a story and a half in height with a chimney in the middle, into which fireplaces in as many rooms as possible were built. Off one end an ell was usually run out, connecting with a woodshed. But the stringing on of barns and other out-buildings beyond the woodshed, so common in northern New England, was rare among the islands. The main part of such a house would perhaps be no more than thirty feet wide and thirty-five feet long. Downstairs there might be three rooms and a kitchen, the latter often in the ell. To make heating easy, the ceilings were as low as six and a half feet. In the center of the house the front door, almost always with sidelights, would open into a square hall, from which a staircase would mount sharply to the second floor, where two or three bedrooms would be tucked under the eaves. Beneath the main part of the house would be a cellar with walls built of slabs hewn from nearby ledges. In front near the granite doorsteps would grow a lilac bush, the symbol of the early American rural home, which would remain, growing to treelike size as the years marched by.

Several of these houses have been bought and "fixed over" by summer people. But one at least is occupied almost in the same manner as during all the one hundred and fifty years of its existence. It squats beneath a high ledge, as gray as its own unpainted clapboards, that protects it from the winter winds but makes a dreary outlook from the back windows. Its solitary occupant is a fisherman of over eighty, the oldest living member of one of the island's oldest families. He has lived in this house only twenty-eight years, he explains. He was married at the age of twenty and his wife died only six years ago. He has a daughter, five or six grandchildren, and two great grandchildren, all living on the mainland. He would live with one or another of them except that he must maintain his own home to receive the full old-age benefit of sixty dollars a month. He doesn't hold much by doctors. When he is sick, he "toughs it out," and so far, though he's had a couple of close grapples with death, he's always "gained the day."

When the electric-power co-op was established, he paid his five dollars membership fee to help the island. But he didn't want the current in the old house. The only modern conveniences that he tolerates are the telephone, which he "couldn't do without," and an oilburning range for cooking and heating, for he is no longer able to cut firewood, which like all old-timers he still considers to be the best fuel. For water he takes a bucket to the well a hundred yards from the house. Back against the ledge is a rickety "little house."

Every autumn, when the nights begin to cool and the sound of the rote comes with the easterly wind across the island, he takes a hammer and drives in all the nails that the frost of the past year has sprung from the narrow clapboards. Around the foundation he piles fir boughs to catch the snow and thus keep out the cold, as has been done every winter for a century and a half.

The inside of the house is simple and bare, like many of the lives that have been lived there. Plain wooden chairs and tables, linoleum floors, a kerosene lamp or two, and the vast spidery bulk of the blackened range are the only furnishings. The old man spends most of his time in a rocking chair by the range, reading his paper and smoking his pipe. Occasionally a summer person or a child visits him and listens to him reminisce. He loves to relate his experiences in mackerel fishing. Foremost among his memories are great past convulsions of nature—the storied *Portland* gale and the fabulous winter of 1917-1918, when the Gulf of Maine must have closely resembled the Arctic Ocean. He is proud of his island and his family. Usually he refers to himself by his given and his family name, seldom by the first person pronoun.

He rarely strays far from his house. But when he does he wanders down the steep pastures to the water's edge where he has a punt. When he has watched the water long enough and dug a few clams, some of which he will drop off at a neighbor's, he returns to his stove and his rocking chair and newspaper. This is an old man's life in a house of which he proudly says, "I don't know but what it's the oldest on Swans Island."

His outlook and his manner of living are not markedly different from those of the early settlers who built the house.

In the middle-third of the last century another type of house was popular on the island. In these the central chimney was replaced by several spindly ones, for this was the age of stoves. They were higher, also, with two full stories. Their main feature was the precipitous pitch of their roofs, almost as steep as church spires, or spruces, after which they may have been modeled in emulation of that tree's snow-shedding powers. Most of the islanders still live in houses of this sort, which date from the generation of the Civil War. They are roomy, ruggedly built, their timbers sometimes fastened by wooden pegs.

The last period of building, which came at the end of the century, represented the lush times of the mackerel. The chief builder at the time was an islander of whom the natives today wrily remark, "He liked big houses." He must have built with the idea that every one of his creations would eventually become a hotel, as several did for a short time. Not only were his houses big, but they were freighted with towers, bulbous projections, pompous mansard effects, and miles of gingerbread. Invariably they were painted green with white trim, colors which the luckless present owners perpetuate in despair of finding any others that wouldn't look even more hideous. Inside, the ceilings are as lofty as those of the drawing room of a Fifth Avenue mansion.

Two monuments to this period are the mammoth halls built by the Redmen and the Odd Fellows. These monstrosities, five stories high, vie with dirigible hangars in bulk. All the Redmen and Odd Fellows in the state of Maine could meet in them and have room to spare. Today the ridgepole and sills of the Redmen's Hall are sagging, the glassless windows gape, and the red paint has faded till it resembles a case of convalescent measles. The selectmen have pronounced the building unsafe for use, but the blow is not great since the Redmen have long since been inactive. The Odd Fellows too have disbanded but they have managed to keep at least half the windows intact in their airy hall. The building serves for town meetings and as a place for dances and itinerant cowboy entertainments. One of the smaller rooms is rented to the town for its office.

In the past fifty years there have been no new houses, with the exception of a few summer camps and two recent suburban-type dwellings made of lumber from an old store. The average age of the houses now occupied on the island must be close to one hundred years.

2

Until 1950 almost every island house was without the conveniences that Americans consider so indispensable. Of the two hundred and ten houses reported in 1950—and this included summer homes—only fourteen had bathrooms and only twenty-two had running water. Central heating was even rarer than toilets. Yet for those aghast at this state of affairs, it should be added that the islanders showed every appearance of enjoying life. This doesn't mean that they were living in blissful squalor. Census reports give lopsided pictures. They indicate the lack of bathrooms and running water, but they fail to record the spotless kitchens, the neat, dustless parlors and bedrooms, and the gleam of nickel trim on shiny black stoves.

The greatest change that home and community life on the island has ever known came with electricity in 1950. A Rural Electrification loan of one hundred and sixty-one thousand dollars was secured to be paid back at the rate of six thousand dollars a year. The killjoys, of course, had their day. The town couldn't afford the debt, the rates would be too high, the generators would burn out because of insufficient power consumption. In short, a little island town had no business aspiring to electric power, even though the rest of the nation had had it for fifty years. But enough people wanted it so that finally a crew "came on" with bull-dozers, cleared out swaths through swamps and over ledges for the lines, and dug the pole holes, or blasted them in solid rock, and strung the wires. Such devastating efficiency had never before been seen. The citizens stood agape as they watched mechanical behemoths flatten whole forests before their eyes. Soon two 155.5 horsepower Diesel motors were pumping out current from the sheet metal shed on the quarry wharf at the harbor. At first only half the families took the current, of which they had to guarantee to use sixty dollars worth a year. But most of the islanders paid the five-dollar membership fee in the co-op and later many of them found money to wire their houses and start using electricity. A few held out—some who could foresee nothing but disaster for this or any other project attempted on the island, some who were too old to try anything new, and some too poor to pay for it. But by now, at least three out of four island houses are electrified.

The coming of electric power coincided with the greatest prosperity the island had ever known—a period when lobsters were plentiful and were selling at inflation prices. To watch a community emerge overnight from the nineteenth century darkness of the kerosene lamp into the glare of the incandescent bulb is a lesson for economists and moralists alike. In a year or so one can observe changes that were spread over half a century of the life of the nation as a whole. It was a change from simplicity to complication, from financial solvency to a multitude of time commitments, from much labor in the form of hand-laundering and sweeping to much worry in the form of unpaid-for washing machines and vacuum cleaners. Storekeepers on the mainland saw the bonanza. The gadgets and conveniences began to pour in, occasionally after full payment in cash, but usually on the purchaser's signing a slip of paper. The potentialities of the island market were staggering. Householders scrapped battery radios for ones operating on alternating current. They tossed out stove-heated irons, foot-operated sewing machines, ice-boxes and kerosene-burning refrigerators, and they toppled privies to make way for bathrooms with water systems powered by electric pumps. Even the few elite who had direct-current home generators put aside their old equipment for the cheaper and more usable current supplied by the co-op.

At the very height of the change-over another bombshell burst upon the island. Bangor seventy miles to the northward was chosen as the site of the first TV station in the state of Maine. At the same time the price of lobsters took a further upward spurt. Word got around; TV salesmen, order books in hand, moved in to do a rousing business. Close at their heels were those older hands at the island business, salesmen of vacuum cleaners and storm windows. The estimate of money spent in 1953 on "improvements" is impressive:

8 artesian wells, averaging nearly 100 feet each, at $8 a foot		$ 7,000
8 bathrooms, pumps, and plumbing jobs to go with the wells		8,000
65 TV sets, averaging $450 apiece		30,000
300 aluminum storm windows	Sold by one salesman in three weeks	9,000
20 aluminum storm doors	Sold by one salesman in three weeks	2,000
30 vacuum cleaners, all sold by one salesman		3,200
Total		$59,000

In the same year the island's income from lobsters, its chief resource, was, according to the Department of Sea and Shore Fisheries, one hundred and seventy-five thousand dollars.

The islanders were quick to find a motive for their expenditures. The reason they give for everything, from building a new school to the installation of a toilet, is that they want their children to have the same advantages the mainland children have. The desire is, of course, unselfish. But it reflects a deep-seated fear of the mainland, a feeling of inferiority towards it, that they don't wish their children to have. Thus

TV, like the radio before it, is hailed as an equalizer of the islanders and the mainlanders. Both sitting in their living rooms see exactly the same pictures on the screen. The islanders make a great point of the "educational" qualities of TV, and when pressed to name an example they cite the circus put on every Saturday morning. "It would cost a family one hundred dollars to go to Bangor to see a real circus. Now our kids can sit in their home and see a circus just like other kids."

Before TV several islanders purchased a movie projector and screen. Once a week movies were given in the Seaside Hall so that island children would not miss this feature of mainland life. At first the apparatus was powered from a portable gasoline generator, but when the Co-op was established, it was run on regular current. Each Saturday evening the island youth, and some adults, would pay a quarter or less to sit through a grade C horse opera or mystery flickering dimly on a fly-specked screen between breakdowns of the projector while the sound track emitted a roar like that of a passing freight train. Barrels of soft drinks and bushels of candy bars were sold at a table to the side, so that the crackling of paper and the rattle of bottles falling to the floor gave variety to the din. Throughout the show the door banged open and shut as swaggering adolescents passed to and fro. Only the young children and the gnarled housewives from the "end roads" made any attempt to follow the action. But shortly after the invasion of TV the movies were discontinued for want of patrons and the projector was turned over to the school to make possible the use of audio-visual aids.

One very tangible advantage is coming from all the TV sets. They are keeping the Co-op solvent and saving its equipment. Up till then not enough current was sold to meet running expenses and to keep the generators from overheating. But each TV set uses an average of three dollars' worth of current a month and also keeps the islanders burning electricity for lighting purposes much later than ever before. Lights shine in island windows till ten or eleven at night. Before TV the black-out was hours earlier.

Another remarkable achievement of the Co-op is not in the realm of physical change. Like the school, it has brought the islanders into a cooperative venture in spite of their individualism and petty division into their three tiny villages. The R. E. A. is now the livest organization on the island. Its annual meeting is a big affair with free suppers at which State University extension workers demonstrate to intensely interested housewives how to cook and keep house by electricity. The whole undertaking has shown the effectiveness of working together. The people are learning that businessmen will consider their problems only when there is a chance of making a profit, as with TV sets and vacuum cleaners. Where there is no chance of profit private enterprise will leave them islanded in the nineteenth century.

3

To the women, electricity with all its accessories has been the greatest boon. The older women used to be virtual troglodytes in their homes. The drudgery of running a household solely by handpower—as most of them did till five years ago—consumed every minute of their time and every ounce of their energy. It twisted their spines, furrowed their faces, and stiffened their joints. The only thing it didn't seem to do was to shorten their lives, which some may have regarded as a questionable advantage. For months they would not leave their houses and backyards, not even for church. A trip to the mainland would be rarer than a birth, a death, or a wedding. Then electricity suddenly unlocked the shackles. The women speak of it as a miracle. They avow they never dreamt it could happen "on here," which to many was a place where nothing ever could or would happen, where life stood paralyzed in awe of the hurried march of mainland affairs.

What the women will do with their emancipation is another thing. The elderly ones are too used to the old bondage to feel quite comfortable when stripped of chains. They remain in their homes, worrying and fussing, making work where none is readily discernible. Freedom for them is a privilege of childhood. After a few years of playing on the beach or in the backyard, a girl should get into harness in the kitchen beside her mother, and perhaps her grandmother. There she should remain till married, when she transfers her labors to another kitchen. A promising girl in a literate family would sometimes be sent to a normal school to study to become a teacher.

For most women the only outside employment either before or after marriage was in the canneries or fish-packing plants. The work was seasonal and poorly paid, but at least it offered an escape from the kitchen. Some women also baited trawl lines during the fishing season—a job suited to the agility and speed of a woman's fingers. But with the specialization in lobsters even these jobs no longer exist. In their place are limited chances to do housework for the few summer people who can afford to hire help. Occasionally a woman picks up some money knitting articles like fishermen's mittens, and one has brought in a little by selling jewelry that she makes from sea shells. Recently a summer person, temporarily trying to make a go of year-round life on the island, has done silk-screen printing of Christmas cards, notepaper, and paper napkins. She herself makes the stencils—views of lighthouses and other coast scenes. An island woman has learned to run off the prints. But paying handicrafts like these are the exception.

The women do much rug-making, embroidery, quilting, and knitting, but most of their output goes to the church sales held every summer.

4

Before the radio the islanders made their own music. In 1918 there were fifty-two pianos on the island, or one for every four houses—a ratio that few communities could equal. In addition were many cottage organs and smaller instruments like the guitar, violin, or accordion. Everyone loved to sing, in church or in groups at someone's home. But with the coming of battery radios to most island homes in the 1920's and '30's there was a rapid decline in singing and instrumental playing. Dancing, too, was once an island pastime, and getting together a band was no trouble in a community where almost everyone played some instrument. One of the public halls or one of the "hotels" would supply the floor. Groups might come from the neighboring islands and there would be great gaiety and mingling of the people. But with the loss of interest in music, dancing fell out of favor. Today an impromptu dance is a great rarity on Swans Island.

On the radio or in what little playing and singing the islanders still do, cowboy and hill-billy music is the favorite. "Home on the Range," "Red River Valley," and "Prince Edward Island" have drowned out the older ballads, many of them of the sea, which can now be found only in such books as Mrs. Fannie Eckstorm's *Minstrelsy of Maine*. Nearby radio cowboy groups often "come on," rent the I.O.O.F. Hall, and put on their program of singing, playing, and folksy talk. A poster like the following will bring out the larger part of the townspeople:

SEPT. 5 COMING SOON

BIG WESTERN JAMBOREE
ODDFELLOWS HALL
SWANS ISLAND, ME.

LITTLE JOE
AND THE COUNTRY BOYS
FEATURED VOCALISTS
BIG SLIM REYNOLDS
CHUCK LITTLE SHIELA PEGGY
(6-year-old Country Girl)

ACCOMPANIED BY
A. TEXAS CROCKETT WITH HIS ACCORDION
BAND AND THE TALKING BANJO

ADMISSION ADULTS 75¢ TAX INCLUDED 35¢ CHILDREN
DOOR OPENS AT 7:00 P.M. SHOW STARTS AT 8:00
AMATEUR CONTEST DANCE

The islanders still cling to two traditional social occasions—the celebrations of Christmas and the Fourth of July. Christmas is observed much as elsewhere except that commercialism is not so evident. By December 1 on the island, preparations are under way for a round of school and church parties. In a land of conifers the people can afford to be meticulous about the choice of a tree. They will consider only the most symmetrical fir; spruce, though shapely, they hold beneath contempt because of its coarser needles. Homemade wreathes and festoons glowing with the fiery red berries of the white alder decorate flower boxes, windows, and front doors. Till a few years ago candles lighted the trees indoors. Now chains of electric bulbs do the job more safely and less picturesquely.

On closing day the school has its long-awaited "Christmas tree," as the whole party including the exchange of presents is called. A day or so later the Girl and Boy Scouts get together for a party. Finally the Sunday Schools have their parties, at which every child receives presents sent by the Maine Sea Coast Mission. The Mission packages always include some candy, something useful such as mittens or a scarf, and a book or some plaything.

Not a person is overlooked on Christmas. The lonely, the down-and-out are not forgotten in the stampede to make the children happy. The shut-ins and the very old get presents from the Mission. Neighbors buy presents and hang them on a tree which they carry to the homes of these unfortunates to cheer them. On Christmas Eve the young people carol them.

On the Sunday before Christmas the churches have their big services. Attendance is larger than at any other time. There is lusty singing of carols and the people give sincere attention to the sermons. On Christmas Day there is no public worship in this rigidly Protestant community. It is a day of family gatherings and present-giving. Nor does the Epiphany mean anything to the islanders. A day or so after Christmas they dismantle the trees and chuck them out. It is all over until next December 1.

The other big day on the island is the Fourth of July, which is suggestive of the rowdyish celebrations that marked the occasion a hundred or more years ago. Things start happening the afternoon or evening of the third. How quickly they progress depends partly on how recently the lobster smack has been in with its cargo of bottled cheer. As the boys—which term for the next twenty-four hours includes all males from sixteen to sixty—apply the bottle to their lips, jalopies begin to stir about the island more hectically and in greater numbers than usual. If there is a dance, as there used to be in the old days, the boys will attend it off and on. But mostly they are on the road. From midnight to dawn the churchbells clang. Every car passing one of the always unlocked churches stops and its occupants take a few pulls at the rope. Noises of rending metal add to the din as mowing machines, road scrapers, and hayrakes are towed at high velocity along the roads.

Hallowe'en on the mainland is panty-waistish by comparison. In the morning fishermen seek their punts in lily ponds half a mile up a hill from the wharves where they tied them the night before. Summer cottagers comb the island for lawn furniture they left out-of-doors in the innocent belief that Hallowe'en was four months away. Ordinary citizens called by nature sadly contemplate back-houses turned upside-down. The sheriff dashes about in his pick-up truck collecting and returning misplaced property and trying to placate townspeople who show spoil-sport tendencies. "You were young yourself once, weren't you? Have you forgotten how boys are?" But the irate householder won't be softened. "They must have been some pretty overgrown boys to carry my privy across the road and heave her into the drink."

This is the closest Maine can approximate Walpurgis Night, and it does it about the right time of the year. So great is the fun that more than one summer person plans his vacation so that it includes an island Fourth with "the boys." It's one good time they would be reluctant to forego.

13

Town Meeting

A MILKY SUN, SLASHED BY THE SAW TEETH OF THE SPRUCES, rises across the harbor on which a film of ice awaits dissolution by the tide and the morning breezes. A motorboat, "going outside" to haul lobster traps, rocks tentatively on the swell at the harbor mouth. But many boats are still at their moorings, where in the changing winds they will pivot all day. This is town meeting day, and already a curl of smoke unwinds from the chimney of the towering Oddfellows' Hall a half mile up from the shore. Soon the selectmen will gather there to "correct the list of voters." At ten o'clock the meeting itself will begin, the 120th in the island's history.

Since 1834 the island has had yearly meetings to elect officers, to levy taxes, to establish schools, and to lay out roads. In 1897 the community was made a town, with added responsibilities particularly in regard to the support of the poor. The plantation meeting of 1896 had rejected by a vote of eighty-two to fifty-nine this town status as provided by special action of the state legislature, for taxes are usually higher in a town than in a plantation. But the next year the legislature passed a mandatory bill, and willy-nilly the island assumed its larger role of self-government and self-support.

On this first day of March, 1954, the townspeople are assembling to consider commitments and enterprises greater than any ever placed before them. In 1839, the first year for which records are extant, the voters "raised" only $265: $90 for schools, $150 for roads, and $25 for administrative expenses. In 1897, the first year of town status, they raised $2,135: $735 for schools, $500 for roads, $200 for support of the poor, and $700 for administration of the government. In 1954 they were to vote $12,001.75 for schools, $1,733 for roads, $500 for support of the poor, $2,400 for a doctor's subsidy, and other sums for such purposes as "advertising the recreational resources of the island" and cleaning up the town dump. The total was to be $24,584.75 for a community of four hundred and fifty persons, or eighty-nine times that of 1839, when the population was about two hundred and fifty, and eleven times that of 1898, when the population was seven hundred and fifty.

This hazy, springlike March morning, as the cars converge on the Oddfellows' Hall, the occupants are unconcerned at the great sum the public weal has come to cost. This is to be largely a young peoples' meeting, though by no means all the young people attend. The fine weather has lured some outside to haul. Lobsters have been scarce and money scarcer. Only fifty or sixty of the town's total of three hundred voters appear. Many of the young, even among those who are present, are indifferent, but most numerous among the stay-aways are the older people, who have been left behind by the race of events in the past ten years and have lost whatever control they have had on local affairs. Recently, according to them, the island has parted its mooring and is adrift in shoal water. They point to the debt of about one hundred and fifty thousand to the R.E.A. for their power plant and to the debt of thirty three thousand to the state for their new school building. Such obligations are a heavy burden on a hundred or so fishermen. And, though they seldom speak of it, at the back of their minds is the spectre of a failure of the lobster supply. Thus many of the old folks sit at home and complain that the town is run by two or three irresponsibles and that the meetings are not fit places for level-headed citizens who believe in paying their bills.

Everybody is acquainted in detail with what has been going on. In each box in the three island post offices has been placed a copy of the "Annual Report of the Municipal Officers," printed at a cost of two hundred dollars at Stonington for distribution to the voters. In its thirty-two pages they have read a complete history of town affairs during the past year: the names of the officers; the valuation of real estate and personal property, with tables giving the numbers of such taxable possessions as boats, radios, musical instruments, horses, oxen, sheep, and cows; the town appropriations at the last meeting with an itemized account of how they were spent; lists of tax delinquents during the past ten years; and a report by the school superintendent, containing a census of scholars, an accounting of the year's expenditures, and an estimate of what next year's budget will be. At the end of the report is a copy of the warrant, duly posted by a constable, and listing the fifty-four articles on which the meeting will have to act.

The islanders save their reports from year to year and build up files of them which they study occasionally in their spare time, with the result that they become expert in town finances and politics. They discuss abatements, overlays, and negotiable notes. By studying the names of those who have been paid for services to the town they can spot favoritism and nepotism. Yet though they see with such a knowing eye, they are reluctant to act on their knowledge and to attend town meeting and oppose measures and practices that they consider harmful.

2

When the meeting opens sometime after ten only thirty-five or so voters are present. Some of these are seated in a little island of rickety chairs in the center of the large second-floor room. Others sit or stand along the walls. Directly before the stage—for this room is the scene of dances and cowboy shows—is a table where sit the three selectmen and the town clerk with his leather-bound record book. On the open floor children run and play and shout. The door at the back is constantly banging as citizens come and go.

The first selectman, a carpenter of benign countenance getting on in years, rises and speaks into a microphone connected with a squawky amplifier, which lifts the voices of the speakers above the din of children playing Indians and women gossiping. Matter-of-factly he calls the meeting to order and requests the nomination of a moderator. At once the moderator of the previous years is elected. He is a young lobster fisherman, a married man with several children who are at present running about with the other youngsters on the floor. He is a man noted for his courtesy, presence of mind, and quiet forcefulness. He doesn't lose his temper even when, during a discussion as to the disposal of the old schoolhouse, some wag suggests putting a match to the building and someone else is silly enough to second the motion. He merely insists on a sensible motion. "You can't burn a public building, even if you aren't going to use it any more." And the sensible motion is soon forthcoming when a young man who owns a truck offers to dismantle the school and cart away the materials if he can have them for his own use free of charge.

The first part of the meeting is given over to the election of officers. First a town clerk is chosen, for he will have to record the minutes. Fifteen or twenty voters scribble a name on scraps of paper and walk with them to the front of the room, where they drop them into a cigar box on the moderator's table. As everyone expects, the man who has been clerk for years is reelected. He is the most energetic man of affairs in the town—a holder of many offices. In addition to being town clerk, he is chairman of the schoolboard, a tireless worker for the better education of the children over the whining and sometimes snarling opposition of the taxpayers. He is also deputy sheriff, collector of excise taxes, assistant sexton, a notary public, a mail carrier, and a member of the school-planning committee, the budget committee, and the public-health and sanitation committee. In private life he used to be one of the two partners controlling the granite quarry and is now half-owner, treasurer, and chief linesman of the Swans Island Telephone and Telegraph Company. He occasionally is the town barber. In his spare time he is a lobster fisherman.

After the moderator and town clerk have been elected, the officers of the past year present their reports, as printed in the booklet previously distributed to every voter. The elections then continue with the choice of a board of selectmen, who will also serve as assessors and as overseers of the poor. The moderator explains that it is customary to elect one selectman from each of the three "sides." The same principle of "one from each side" obtains in the choosing of a school committee and constables. This year the previous selectmen are retained, but their compensations are increased. The first selectman will now make ten dollars a day; the second, nine; and the third, eight. The year before each worked about fifty days on town affairs.

Next the attractive young wife of a local storekeeper is elected town treasurer, a position in which she has already served for four years. She is the first woman in the town's history to hold an important office. Since she was first elected, there has been a woman tax collector, and several women have sat on various committees. The town has yet to elect a selectwoman, but the possibility is no longer discounted. Until now the men have monopolized law enforcement and politics. The holders of important positions—selectmen, town clerk, treasurer, head of the school board—have tended to become fixtures. Twenty or thirty years for one man in such an office are not unusual. The only major position to change hands often is that of road commissioner. The "road patrol," as he is called, is paid well, but he is open to criticism. Imperfections in road surfaces are more noticeable than those in fiscal or school policy.

The election hits a snag, at this meeting, when the time comes to choose constables. Since law enforcement is almost entirely relegated to the deputy sheriff, the most exciting thing a constable does is to tack up warrants of town meetings in the three post offices. As for wages, he is lucky if he makes ten dollars a year. Consequently, the first to be elected from the Atlantic side declines the honor. He is a lobsterman with four-hundred traps out. He is about to start a lobster-buying business of his own. He clearly will have no time even for the limited duties of a constable. Next a man who is not present is elected. A willing candidate is found for Minturn, but since no one from the village of Swans Island is available, a tradition is broken by the election of a second man from Atlantic, with the understanding that he is to represent the other village.

Other dignitaries are chosen—two sextons, a collector of excise taxes, a surveyor of wood and lumber, and a fire warden. As each officer is picked, the town clerk swears him in on the spot. Only a half dozen or so vote for some of these positions. In the election of the fire warden, only one vote is cast, but on being asked by the moderator, the assembly declares itself satisfied. The election of the collector of personal and real estate taxes fills the last political office. Of the past year's taxes $1,876.32 remain ungathered, so that a new collector seems desirable. A man who has held the job in the past with great efficiency is finally chosen. He acknowledges the honor and announces that if the fishermen have any money next fall he will get the taxes. His terms are three percent, part of which he will turn over to the school fund. The previous collector turned over all her percentage to the school.

Taxes, of course, are a sore point throughout the meeting. One irate citizen says that the owner—or previous owner—of a house on which there has been a tax lien has been renting an apartment in it and keeping the rent. The selectmen reluctantly admit that this had been so. They had winked at the situation because they thought the owner "could use the money." They promise not to make a practice of that sort of thing. A special motion is then passed authorizing the selectmen "to sell and dispose of any real estate acquired by the town for non-payment of taxes thereon and to execute quit-claim deeds for such property." It is also voted that the town will collect 6 percent on taxes not paid by October 1—a date chosen because usually at that time of year the lobsters "are coming well" and the fishermen have as much money as they ever have.

At noon the meeting adjourns for an hour. The last piece of morning business was to "raise" $8,603 for school expenses—teachers' and superintendent's pay, supplies, and textbooks. The people have already started to straggle downstairs where for one dollar a plate the Rebeccahs have prepared a dinner of beans, brown bread, salad, macaroni, coffee and pie. There have been no hard words upstairs, and there won't be. Anyone who wished to say something—even if he was not a legal voter—has had his chance to speak. Most, in fact, have said nothing but have participated by voting for officers on slips of paper or by show of hands on the issues brought up. Now all sit together at the long tables and eat, but only after the minister has said grace. A separate table has been spread for the children, for school has been closed so that the teachers may attend the meeting in their own home town on the mainland.

3

The meal is leisurely. Nothing in the day's proceedings is done with haste. But by quarter past one the officers are again at their places upstairs and the townspeople have begun to drift up. The afternoon's work will be almost entirely the appropriating of money. The men are more alert now. A shrewd, shifty look creeps over the faces of some of them. They gaze at the floor and glance out of the corners of their eyes as each sum is mentioned: $300 for hot lunches at the school, $1,285 towards the payment on the school building, two hundred dollars for a garage for the town truck and school bus, eight hundred for town roads and bridges. Someone wants to know why this last sum is so high. Various people speak, among them a woman from a shanty at the North, who says that there is no longer any road out her way. The moderator corroborates her; the road, in fact, was dug up and, as gravel, taken elsewhere for fill. On the strength of such evidence the sum of eight hundred dollars stands.

The recommended appropriations have all been figured out ahead of time by a budget committee of five from each side, and their recommendations are followed without demur in the meeting. The town clerk is proud of this device, which saves much time and argument. What wrangles do occur are good-natured ones over small matters. For example, the question of the public wharf—formerly the steamboat landing—consumes almost an hour. The wharf was bought by the town three or four years ago for the sum of two thousand dollars, but nothing has been done to keep it in repair. Large-sized draggers coming into the harbor for the night tied up to the spilings and gradually broke them loose. The wharf, in a state of dangerous delapidation, is now used by a half dozen lobstermen as a storage place for their traps. At the town meeting someone fears that if one of these men falls through the rotten planking and breaks a leg, he will sue the town. The moderator asks if that wouldn't be a pretty cheap thing for anyone to do who is getting his facilities free, but the meeting decides that the selectmen should place a sign on the wharf disclaiming any legal responsibility. As for mending the wharf, no one wants the town to put out any money on it. Someone suggests that the town charge rent for its use, but the question of who would collect the rent is unsolvable. Finally it is voted that if the six present users wish to make some repairs at their own expense and risk, the town would place no obstacles in their way. Also a certain man who wishes to establish a fish-buying business on the island may build a shack on the wharf and make other improvements, if he so desires. Several citizens hope that Mr. Bousfield of the Maine Sea Coast Mission will take over the wharf as a place for the Mission boat to tie up. He would keep it in good repair and, of course, permit anyone else to use it.

Another discussion develops over the town dump. The present one has been condemned as unsightly and a breeder of rats. One hundred dollars are needed—and voted—to clear it up. But some other site must be found. The moderator speaks sternly of the litter made by people dumping their trash on the beaches, particularly near the homes of some of the older summer people. "It is disgraceful that that should be done to people who have been coming here so long and admire Swans Island the way they do." No one takes issue with his remarks. These are lovers of neatness, the possessors of spotless kitchens and boats. The indiscriminate disposal of garbage offends them more deeply than it does the summer people. Finally, a committee is appointed to study the matter and make some recommendations.

The meeting draws on towards the fifty-fourth and last

article on the warrant. There has been little talking for the sake of talking. Yet anyone may speak. At no time does the moderator attempt to railroad an issue through. Always he is saying, "It's your decision. What do you think?" The only tense moment during the day was when someone questioned how the town could spend $876 removing two light falls of snow from 12.85 miles of road. The answer is that the state contributes sixty-five dollars per mile of road for snow removal; it would be folly not to use it.

The last article on the warrant now comes up: "To see if the town will vote to change the name of Finney (Phinney) Island, so-called, to Asa's Island as requested by Asa Phillips, Jr." The moderator's eyelids narrow as he reads the proposition. This island, an acre or so in size, is just north of the main island—named after an enterprising Irish citizen of the last century. But, the moderator explains, it is a hobby among yachtsmen to have an island namesake. Grins form on the lips of the townspeople and someone mutters the cliché, "These summer people do the craziest things." Mr. Asa Phillips, the moderator continues, is a prominent lawyer. His attorney approached some island dignitaries concerning the change of name. They had been non-committal; they had said that the islanders always try to be obliging. But they thought that Mr. Phillips might be interested to know that the town was currently under heavy expense in building a school and everything wasn't covered by the state loan. Mr. Phillips is a public-spirited man and not one to receive favors without proper show of gratitude. "Now," the moderator went on, "we've received nothing yet, but neither have we given anything. For my part, I see no reason why we need to call the island anything but Finney ourselves. But if we put this through, the name will be Asa on the charts when they are reprinted." For the town, the moderator saw in the proposition a sum well in three figures. The vote is unanimously in the affirmative.

A few other items of business not on the warrant are discussed and the meeting adjourns. During the afternoon a mist has crept in, and a keen wind has begun to blow off the ocean to the eastward. In the twilight the people drive home hoping that lobsters will become more numerous before the tax deadline next October 1. But they have a feeling of fulfillment, of a job done to the best of their abilities and in accordance with the tradition of their island and race. There are no flamboyant mouthings about democracy any more than there was talk for the sake of talk in the meeting itself. Nor is there any smug condescension and regret over those who stayed at home out of inertia or spite. Compulsory attendance at town meetings has never been a part of the tradition.